D0758777

PLANT LIFE OF THE QUATERNARY COLD STAGES: EVIDENCE FROM THE BRITISH ISLES

The Quaternary period is characterised by extensive glaciations in the Northern Hemisphere, separated by much shòrter temperate stages. For Britain and Ireland, the vegetational history of the temperate stages is relatively well known, but the flora of the cold stages has never been considered in any detail, despite the fact that records of pollen and macroscopic plant remains have accumulated over the years. In this book, Richard West brings together for the first time the published information on the Quaternary cold stage flora of over 80 sites in Britain and Ireland to present a factual cold stage flora from the fossil record. His account provides a basis for an interpretation of the flora, vegetation and environments of some of the most extraordinary periods in the Earth's most recent history, now only seen in the imperfect mirror of today's Arctic, and which precede the life we see today. This important study aims to reveal the nature of an environment, relatively stable, but totally different from that of today. As such it will be significant not only to those interested in the Quaternary, but also to a wider audience of those studying the present flora, fauna and environment, including climate and climatic change.

Includes a database of the cold stage flora in a comma-delimited format.

RICHARD WEST F.R.S. is Emeritus Professor of Botany at the University of Cambridge and Fellow of Clare College, Cambridge. His research career at Cambridge spans nearly five decades, during which time he has made a significant contribution to the field of Quaternary research. He is author of *Pleistocene Geology and Biology* (1968, 1977), *The Ice Age in Britain* (1972 with B.W. Sparks), *Preglacial Pleistocene of the Norfolk and Suffolk Coasts* (1980) and *Pleistocene Palaeoecology of Central Norfolk* (1991).

PLANT LIFE OF THE QUATERNARY COLD STAGES
EVIDENCE FROM THE BRITISH ISLES

by
R.G. West
Emeritus Professor of Botany, University of Cambridge
Fellow of Clare College

CAMBRIDGE
UNIVERSITY PRESS

PUBLISHED BY THE PRESS SYNDICATE OF THE UNIVERSITY OF CAMBRIDGE
The Pitt Building, Trumpington Street, Cambridge, United Kingdom

CAMBRIDGE UNIVERSITY PRESS
The Edinburgh Building, Cambridge CB2 2RU, UK www.cup.cam.ac.uk
40 West 20th Street, New York, NY 10011–4211, USA www.cup.org
10 Stamford Road, Oakleigh, Melbourne 3166, Australia
Ruiz de Alarcón 13, 28014 Madrid, Spain

© Cambridge University Press 2000

This book is in copyright. Subject to statutory exception
and to the provisions of relevant collective licensing agreements,
no reproduction of any part may take place without
the written permission of Cambridge University Press.

First published 2000

Printed in the United Kingdom at the University Press, Cambridge

Typeface Times NR 10/13pt.   *System* QuarkXPress®   [SE]

A catalogue record for this book is available from the British Library

Library of Congress Cataloging in Publication data

West, R.G.
    Plant life of the Quaternary cold stages: evidence from the
British Isles / by R.G. West.
        p. cm.
    Includes bibliographical references.
    ISBN 0 521 59397 2 (hb)
        1. Paleobotany – Pleistocene.   2. Plants, Fossil – Great Britain.
3. Plants, Fossil – Ireland.   I. Title.
QE931.2.W47   2000
561′.1941–dc21   99-26459   CIP

ISBN 0 521 59397 2 hardback

1.2
47
000

For Hazel

Contents

Preface

Those studying the Earth's biota will realise that to understand present life on earth we have to know much about the antecedent life. That knowledge, of life in times past, must rest on a geological foundation. Immediately a biologist departs from the present to the past, many aspects of earth science have to be enrolled in our investigations – including geomorphology, stratigraphy, sedimentology and chronology. This is a major challenge to a biologist. The reverse also applies, that it is a major challenge for a geologist to get involved in the study of past biota, which will include taxonomy, morphology, ecology and physiology.

Such combination of scientific disciplines is a characteristic and necessity of research into the life and environments of the last few million years, the Quaternary Era, with its many climatic changes of alternating temperate and cold stages. The combination is certainly a challenge for those wanting to understand the complex interactions of life and environment in the geologically recent past.

There is an equally daunting challenge in the understanding of the plethora of data on these matters which has accumulated in the last forty or more years. This accumulation has more recently been very much encouraged by the wider perception that climates change, that climate change affects life and environments, that the climate system has to be analysed, and that recent past changes of climate must be understood to assist in providing powers of prediction.

In the lifetime of my own research the application of specialisms old and new in biology and geology to Quaternary research has increased enormously, in parallel with the accumulation of data. It now seems hardly possible that a single researcher can take into account the many facets of the sciences involved and the data produced.

As an undergraduate I read both botany and geology and I then had a decision to take in deciding which of these two took priority of interest. The answer lay in Quaternary research, a subject then rapidly developing in Cambridge under the stimulating leadership of Sir Harry Godwin, Director of the newly-formed Subdepartment of Quaternary Research in the Botany Department. My own research has covered some geological aspects, including stratigraphy, sedimentology and periglacial matters, and some biological aspects, including mainly the forest history of temperate stages in the British Quaternary, studying both pollen assemblages and macroscopic plant remains. But I have also studied floras from the cold stages, and latterly became more interested in them because of their peculiarities. In trying to understand them I have made excursions to Svalbard, the Canadian Arctic Archipelago, Yukon, North West Territories and the arctic slope of Alaska. I realised that there has been no attempt to bring together the large amount of data on cold stage floras of the British Isles (or elsewhere), so that a coherent view of the flora and vegetation of those times could be presented. Cold stages of the Quaternary occupy a major part of Quaternary time, and their importance in the study of past biota and environments can hardly be underestimated.

This account of cold stage floras and vegetation tries to remedy this situation. No doubt there are omissions of fact and interpretation – it is a vast subject. At least, I hope it will provide a basis for an interpretation of the flora, vegetation and environments of extraordinary periods in the earth's recent history, now only seen in the imperfect mirror of the present-day Arctic, and which precede the life we see today.

But I also hope that this work will set a scene not only for those interested in the Quaternary, but also, just as important, for the wider audience of those interested in the present flora, fauna and environment, including climate. I hope the study will reveal the nature of an environment, relatively stable, but totally different to that of today, which existed over thousands of years at times in the Quaternary, and which only faded away some 13,000 years ago as climates improved (from the point of view of today's north temperate biota).

R.G.W.
December 1998

Acknowledgements

I am indebted to many colleagues for their assistance and support in the compilation and completion of this work. The project started in 1990 with the acquisition of the fossil records, a task which continued for some years and was only possible through the assistance of Mary Pettit, who worked on the database to 1997. I am grateful to her for this fundamental contribution, and to The New Phytologist Trust and the Natural Environment Research Council for grant-aiding the work of compilation.

Unpublished records have been contributed by colleagues, including C.A. and J.H. Dickson, Cunhai Gao, P.L. Gibbard, A.R. Hall, Mary Pettit, J.D. Scourse and P.F. Whitehead. A.H. Fitter provided information from the Ecological Flora Database. S. Boreham, M.H. Field, R.D. Meikle, M.E. Pettit, C.D. Preston and R. Newnham advised on matters relating to the database, identification and plant distribution. I have received help from many colleagues more experienced than myself in northern matters, both in terms of their knowledge and in matters related to field visits, including C.R. Burn, J.J. Donner, M.E. Edwards, C.R. Harington, R.E. Nelson, and C.E. Schweger. All these I thank for their varied and helpful contributions. Finally, I acknowledge with gratitude the great debt to my wife, Hazel, for her strong and continued support over the years.

1

Introduction

Palaeoecology aims to study the ecology of past times: how to work out, from the geological record, the relation between plants and animals and their environment at particular times in the past. Palaeoecology gives the fourth dimension of time to ecology, and the necessity of its study for understanding present-day plant and animal communities and how they came about is obvious. Yet many books on ecology take only a present-day view of life on earth. They may stress the importance of the study of the dynamics of communities in the short-term view imposed by human life, but much less so the longer term view derived from the geological record. The recent and widespread recognition of global climatic change has now placed palaeoecology in a very significant position, able to provide evidence of past climates and of the biota which have preceded today's flora and fauna. So they provide information on how changing climates might affect today's flora and fauna.

Our present geological era, the Quaternary Era, began over a million years ago, and is characterised by marked cyclic climatic change, in contrast to the rather more stable climates of the Tertiary. The nature of this climatic change varies from region to region, as does climate naturally today. In north-west Europe, the area that concerns us, the changes are seen as alternations of temperate climate with forest vegetation and cold or severe climate with a lack of trees and presence of herb vegetation. These broad vegetation types thus characterise temperate and cold stages of the Quaternary, with cold stages occupying the major part of the time. At present we are in a temperate stage (post-glacial, Holocene Epoch or Flandrian Stage), which began only some 10,000 years ago, and succeeded a cold stage (Devensian Stage) which lasted at least 100,000 years. Unlike those of the earlier temperate stages, the post-glacial forests have been decimated by agricultural activities in the last 5000 years.

The great ice advances of the Quaternary, the reason for giving the period the epithet 'Great Ice Age', took place during the cold stages. Each ice advance did not occupy the total time of the cold stage in which it occurred, and there may have been more than one ice advance in each cold stage. Within each cold stage there may also have been periods of climatic amelioration, as well as the variations which led to ice advances at particular times. Cold stage climates were evidently complex, in apparent contrast to the temperate stages, which show the invasion and development of forest vegetation. For the reason that cold stages are characterised by periods of glaciation, they have been called glaciations or glacial stages, with the intervening temperate periods termed interglacial stages. Such points of nomenclature are discussed later.

Since the food chain for our biota starts with plants, it follows that the study of past floras and plant communities is absolutely essential as a basis for understanding past biota as a whole. Investigation of flora and vegetation of the cold (or so-called glacial) stages, which occupy a major part of Quaternary time, thus provides a necessary background for understanding the faunas of these times of what we would call severe climate, as well as how the flora and fauna relate to those of the present day.

The identification of Quaternary plant remains began to be taken seriously in the latter half of the last century by palaeobotanists mainly working on the floras of older rocks. The early investigators in north-west Europe, such as A.G. Nathorst, N. Hartz and C. Reid, were concerned with the identification of macroscopic plant remains in both temperate and cold stage sediments, at a time when the knowledge of Quaternary stratigraphy was in its infancy. In the early part of this century analysis of pollen and spores (palynology) developed as a technique, and its potential as a means of investigating vegetational history in the Quaternary was immediately recognised, to a degree eclipsing the study of macroscopic plant remains. Godwin (1968) has described the history of this application of palynology to Quaternary vegetational history in the British Isles.

A very large number of palynological studies of vegetational history of the last 10,000–13,000 years in our area has been published, covering the latest part (late-glacial) of the last cold stage (Devensian Stage) and its replacement 10,000 years ago by the present temperate stage (Flandrian Stage) (Birks 1996; Chambers 1996; Greig 1996; Mitchell *et al.* 1996). The European Pollen Data Base has been compiled to make these data accessible (de Beaulieu 1996). The result is a very detailed knowledge of vegetational history in north-west Europe in this period, with the study of events and correlations facilitated by the application of radiocarbon dating.

Thus it has been possible to construct maps showing the timing of the migration of forest trees and their expansion, expressed by isopolls (lines of equal pollen representation) for particular pollen taxa, often genera in the case of trees (e.g. Huntley & Birks 1983). Using these changes in distribution in time of particular trees, together with estimations of the climatic controls of the same tree at the present time and the relation between present-day forest composition and pollen deposition, transfer functions (climatic response surfaces) have been used to reconstruct climatic change in the past (Huntley 1993).

In the cold stages, however, we are concerned with long periods of time in which vegetation of low stature, mainly herbaceous and locally very variable in character, was prevalent. Changes in mainly herbaceous pollen spectra arising from such vegetation are far more difficult to interpret than changes associated with forest history. In addition, the vegetation cannot necessarily be paralleled at present, as discussed later. Many analyses of pollen deposition in areas of present-day tundra vegetation have been made and related to the parent plant formation (e.g. Ritchie *et al.* 1987). These have been used to attempt reconstruction of cold stage vegetation and palaeoclimates, as with the use of the forest pollen analyses described above (Guiot 1990; Seret *et al.* 1992). These attempts, however, are more difficult with the cold stage pollen data for two reasons. Modern analogues are difficult or impossible to find, and the taxa represented in the fossil assemblages can usually be only identified to families or genera, each of which may be rich in species of varying environmental requirements. Records of macroscopic plant remains, more readily identifiable to a specific level, would enhance such approaches to past plant communities and climates in cold stages.

It is the possibility of combining the evidence for past floras from pollen and macroscopic identifications which gives an added dimension to the study of cold stage floras, a dimension which has not generally been exploited in the study of forest history of temperate stages or of long records of vegetational history from deep lakes.

The compilation of a cold stage flora database of higher plants for the British Isles is an attempt to provide a basis for an improved interpretation of cold stage vegetation and climates, using records of both pollen and macroscopic plant remains. Each of these two categories of fossils makes its own significant but different contribution to the interpretation, both in regard to the constitution of the vegetation and to the climatic interpretation.

These are the kind of questions we might wish to be answerable from such a database:

In regard to flora and vegetation:

> What taxa grew where and when, and how abundantly?
>
> What can be said about the representation of taxa as pollen or macro remains or both?
>
> What is the present distribution of species found in cold stages?
>
> What is the nature of the flora as a whole, in terms of biological properties, such as life form, life span, variation?
>
> How far can plant communities be identified in the fossil record (a much more complex problem than with forest history in temperate stages)?
>
> What species are found in both cold and temperate stages (i.e. appear to have climatic tolerance)?

In regard to taphonomy:

> What are the complexities of fossil input into cold stage sediments and how do these affect interpretation of the fossil assemblages?

In regard to environment:

> What can be said about climate and soil conditions from the plant fossil record?
>
> Can variations in cold stage climate be detected in the plant fossil record or can they be expected to be detected?

The database should also make the cold stage fossil record much more easily accessible, for example, to those interested in proxy data for climatic reconstruction, phytogeographers who want to know the history of a species, taxonomists who are involved in working out variation in species, and palaeontologists who are interested in geological and evolutionary aspects of the subject. The analysis and discussion to follow are intended to illustrate principles and problems of the interpretation of the cold stage flora. Further analyses may be made by specialists, and data can be added as new sites are investigated. The richness of the fossil record may not have been widely appreciated. Nevertheless, the corpus of records must be regarded as very incomplete. Many more data are required for the study of regional vegetation and climates across the British Isles and the north-west European continent.

Consideration of the data in the following chapters is not meant to be an essay on arctic floras and vegetation. Rather it is meant to relate cold stage floras and vegetation to what is presently seen in northern lands, and to underline the problems which beset such a comparison. A perspective is taken which views the present arctic flora as derivative from the cold stage flora, a hardy remnant of the more southern and widespread cold stage flora of the Quaternary, which could spread far north as climate ameliorated in the temperate stages, leaving behind other species which could accommodate themselves to temperate conditions or to suitable refuges in mountains or elsewhere.

But before describing the database in more detail, it will be useful to give a brief historical survey of the subject, and to discuss two aspects of the study which can give rise to problems, such as the definition of terms and the question of analogues.

A brief historical survey

The development of phytogeography in the early part of the nineteenth century corresponded with increasing knowledge of recent geology. Edward Forbes (1846) realised the significance of the postulation of a widespread 'glacial sea', demonstrated by the occurrence of 'boulder clays and pleistocene drifts', for the presence of northern species of plants in the British Flora. He supposed that there was a spread of these plants around the margin of the 'glacial sea', but their distribution later became restricted to elevated mountain regions as climate ameliorated and the bed of the 'glacial sea' was uplifted. Darwin, in the *Origin of Species* (1859), devoted a lengthy section of his book to 'Disperal during the Glacial period', and largely followed Forbes' ideas in explaining the disjunct distribution of northern plants in the European mountains.

Later in the century, the Quaternary plant fossil record, termed by Godwin (1975) the factual basis for phytogeography, started to enlarge. For example, A.G. Nathorst, the Swedish 'hardrock' palaeobotanist, began to analyse (Late) Quaternary assemblages of macroscopic plant remains, which contained what he termed 'glacialpflanzen' or 'arktische pflanzenreste' (e.g. Nathorst 1914). Clement Reid (1899), in his book on the origin of the British Flora, identified many such plants from sites in Britain. As a result of these developments, the relation of phytogeography to the fossil record became a focus of study.

In 1936, a discussion was held at the Royal Society on the 'Origin and relationship of the British Flora'. A.C. Seward (1935), who introduced the discussion, referred to the controversial question of the effect of the Ice Age upon the plant world. Questions were raised about the survival of the British flora during the glacial periods and the relation of the present distribution of northern plants to climatic change: for example, whether the flora had been largely obliterated in the glacial stages (the *tabula rasa* hypothesis;

Reid 1911) or whether there had been perglacial survival on nunataks (Wilmott 1935).

By this time pollen analysis had become well established, complementing knowledge gained from macroscopic plant remains. But the state of Quaternary stratigraphy was much less well known, with far less information about the number of ice advances, the number of temperate (interglacial) stages and the detailed stratigraphy of the cold (glacial) stages in the glaciated and periglacial areas.

In recent years knowledge in all these areas has greatly increased, as has the fossil record. Since Bell's detailed account of last cold stage (Devensian) floras in Britain (Bell 1970) and Godwin's (1975) listing of fossil records to 1970, many more floras from the Devensian and earlier cold stages have been analysed. Godwin (1975) discussed the record and elaborated the relationship between phytogeography and stratigraphy in his classic book on the history of the British Flora. Dickson (1973) has treated the bryophytes likewise. Since that time much further knowledge has accrued, not only of fossils and stratigraphy, but also about present northern vegetation, proxy data for palaeoclimatology, interpretation of fossil assemblages in terms of present vegetation, the taphonomy of fossil assemblages, and absolute dating, especially radiocarbon dating. All these aspects have to be considered in relation to our interpretation of the fossil record, making the task of interpreting cold stage floras very different from the original simple and original 'Forbesian' statement about the relation between phytogeography and the 'glacial sea'.

The connection between geologically recent cold climates and the origin of the British Flora is one of the first areas in which phytogeography came to be seen as clearly related to geology. From this historical point of view, the present study is a development of the same thesis, attempting to take into account the vast geological and biological advances which have been made to the present. What is presented here may guide interest and activities in future studies, giving an idea of the problems and identification possibilities via pollen and macroscopic remains, and for interpreting cold stage environments, flora and fauna.

Definitions

It will be useful to discuss the definitions of some terms commonly used in Quaternary palaeoecology, since confusion often arises from their varied usage. The terms concern climate and vegetation, as at present and in the past. The problem is that the terms concerned naturally arise from what is seen at present. So there is a consequence that terms derived from present conditions are used to characterise past conditions, which may lead to incorrect conclusions, forcing past climates and vegetation into the likeness of present-day entities. With this problem in mind, we can first consider definition of present climatic conditions and vegetation.

Present climate: arctic and subarctic

A great variety of definitions have been proposed for these terms, which are used both capitalised and adjectivally without capitals. Polunin (1951) commented:

A bad blot in the literature is the persistent vagueness surrounding the use of the term 'arctic', whether it be employed in the adjectival form . . . or as a substantive implying a region, viz., the Arctic. It is common for an author to term a plant (or its range) arctic when it reaches an area which according to his conception (or mere copying) constitutes part of the Arctic; but what this last is, or where it begins or ends, evidently varies greatly in different authors' minds.

Any student of the matter will agree with Polunin's sentiments. Generally, 'Arctic' is used to define a region, while 'arctic' refers to conditions associated with such a region. If required, a reasonable and clear definition of the terms as regional terrestrial entities, based on climate, would seem to be the definitions of Young (1989), based on summer temperatures, which are as follows:

Arctic – mean temperature of all months less than 10 °C, with at least one month below below freezing.
Subarctic – mean temperatures of no more than four months above 10 °C, and one or more below freezing.

Present vegetation

The Arctic is characterised by a lack of tree growth, in contrast to the coniferous forest of the Subarctic. The simple treeless character of the Arctic is balanced by the vast variety of vegetation which occurs in the area, promoted by the lack of a forest canopy. It has been divided into areas of High, Middle and Low Arctic on the basis of **bioclimate**, using differences in temperature and the length of growing season for plants, differentiating such features as plant cover percentage, the variety of herbaceous vegetation and the presence of shrub vegetation. Edlund (1987) and Ritchie (1987) have described the bioclimates of the Canadian north, the latter providing a range of climate diagrams for the diversity of conditions from temperate to arctic.

Tundra is the term most commonly used to describe the vegetation of the

Arctic. It is helpful to quote the *Shorter Oxford English Dictionary* (1947) on this term:

1841.(Lapp.) One of the vast, nearly level, treeless regions which make up the greater part of the north of Russia, resembling the *steppes*, but with Arctic climate and vegetation. Also applied to similar regions in Siberia and Alaska.

As will be discussed later in relation to the interpretation of cold stage floras and vegetation, the treelessness has been associated with low annual carbon gain of the vegetation, low soil temperatures and winter desiccation.

It is important to realise the complexities of the many vegetation units and plant communities of the Arctic, described by numerous authors (e.g. Batzli 1980; Andreev & Aleksandrova 1981; Bliss 1981a). Young (1989) has distinguished broad divisions as follows: sedge meadows, tussock tundra, wet tundra, mesic tundra, shrub tundra, fell fields, polar steppe and polar desert. Chapin and Shaver (1985) have described characteristics of major Arctic vegetation types in north America: the high Arctic polar desert and polar semidesert, and tundras typified as wet sedge-moss, tussock, low shrub, tall shrub and heath. In general terms, the shrubby tundras occur at lower Arctic latitudes, the other tundra types more frequently in the mid and high Arctic. The complexity of the vegetation is in contrast to the relative paucity of the Arctic flora. Young (1971), who divided the Arctic into four floristic rather than vegetation zones, commented that little more than a thousand species of vascular plants occur in the Arctic.

Communities with abundant grasses occur widely in the Arctic. If they were not in the Arctic they might be termed steppe communities. Nevertheless, polar steppe is a recognised vegetation type in the tundra. Grassland (see Bliss 1975) is a term which can be properly used to describe the floristic affinities of such vegetation, and which avoids a temperature connotation, though it does imply a continentality or aridity of climate, insufficient for tree growth, which may occur in the Arctic region (polar steppe) or to the south (steppe, prairie). The term steppe-tundra then signifies climatic conditions which promote grassland, but allow the growth of species nowadays considered Arctic or steppe, i.e. variation along gradients of annual and seasonal temperature and precipitation. If we find evidence for such communities in the cold stage flora, the ascription to a particular climate type must then be cautious.

Past climate

Since there are no measurements of past climate available, we have to use proxy data for their reconstruction. There is a wide array of geological and

Glacigenous or periglacial formations

Interstadial climatic development	Interglacial climatic development
	Arctic
Arctic	Subarctic
Subarctic	Boreal
Boreal; summer temepratures essentially lower than the climatic optimum of the post-glacial period	Temperate climate with a summer temperature at least as high as during the post-glacial climatic optimum of the area in question
Subarctic	Boreal
Arctic	Subarctic
	Arctic

Glacigenous formations

Figure 1.1. Climatic development of interstadials and interglacials according to Jessen & Milthers (1928).

biological sources of such proxy data, listed by Bradley (1985). We are concerned here particularly with the use of palaeobotanical data for the recognition of broad climatic events in the Quaternary. Jessen and Milthers (1928) distinguished **interglacial** events from **interstadial** events by the degree of climatic amelioration, determined from palaeobotanical data. Their scheme is shown in Figure 1.1. The terms, as used by Jessen and Milthers, might well be put in a category of **palaeobioclimatic** terms, identifying past climates from biological data. This usage would parallel the use of the term bioclimate to characterise present terrestrial climates, as with the subdivisions of the Arctic mentioned above.

In considering cold stage floras we are concerned with stadial and interstadial events. The 'stadial' terminology developed originally as a result of the subdivision of glacial stages into stadia, associated with periods of ice advance (see Nilsson 1983). However, with an increased knowledge of the periglacial environment and its history, it appears that glacial advance is not necessarily a concomitant to stadial periglacial conditions. This problem is avoided if we use the term pleniglacial, used by Van der Hammen (1951) to describe the non-interstadial parts of the Netherlands periglacial succession in the Tubantian (Last) Cold Stage of the Netherlands, prior to the Late-Glacial. Full-glacial is an alternative version of pleniglacial. Where the stadia of ice advances fit into the periglacial–pleniglacial succession is an important question, because it implies that pleniglacial climates are far from uniform.

In this account of the cold stage flora, the Jessen & Milthers (1928) definition of interstadial is used, based on degree of amelioration of summer temperature indicated by forest development. But it should be remembered that the presence of forest is also controlled by seasonal precipitation as well as temperature. The Early Devensian Chelford Interstadial is identified as an interstadial on the basis of the develpment of forest. 'Interstadial' is not used to identify geological events related to changes of sediment type, such as a bed of peat sandwiched between cover sands, which of itself need give no clear evidence of climatic change, though of course what may be termed **palaeolithoclimatic** units such as cover sand occur abundantly in the Quaternary. Thus the clay bed in the Corton Sands, which lies between two tills of Anglian age in East Anglia, is not interstadial using the Jessen & Milthers criteria. A most interesting problem for a palaeobioclimatic definition of interstadial arises when biological evidence for climatic amelioration taken from palaeobotanical and palaeozoological evidence is at variance, as occurs in the Middle Devensian Upton Warren Interstadial Complex. This very significant matter is discussed in Chapter 13.

As examples of the problems of definition we can refer to the use of 'Arctic' to describe plant beds, where fossil assemblages have been found to contain plants of arctic distribution, e.g. the Lea Valley Arctic Bed. As Wilmott (1935) pointed out, these assemblages may contain many other species which are now associated with temperate climates, and he questioned the use of the term 'Arctic'. The description of vegetation as tundra at lower latitudes in pleniglacial times is also questionable, in view of the vast difference in bioclimate between present Arctic tundra areas and lower latitudes.

Analogues

In the interpretation of cold stage floras and faunas, the search for present-day analogues has always presented an important problem. Are there indeed any present-day analogues? This question has long been discussed, especially in relation to tundra and steppe; for example, by Nehring (1890) in his classic *Ueber Tundren und Steppe der Jetzt- und Vorzeit* and by Grichuk & Grichuk (1960) in their work on Russian periglacial floras, in which they postulated tundra-steppe communities at the margin of ice sheets.

A discussion of present-day analogues for cold stage floras is essential for their understanding. Two elements of the discussion are involved. The first, and fundamental, one is the physical environmental element; the second, and derivative, one is the biological element, dependent on the characters of the first.

In the wider view, it is now well understood that the physical environmental element undergoes continuous change against the background of changing solar radiation patterns (Berger *et al.* 1984), leading to the possibility of no-analogue situations (Huntley 1996). But we have to work to a finer focus, and in interpreting cold stage floras our thoughts of analogues naturally turn to areas where severe cold conditions obtain today in higher latitudes or altitudes. The question then becomes 'are there physical environmental analogues for cold stage conditions in the middle latitudes we are considering?'

The significant characters of possible analogue areas concern the plant growth environment, climate and geology. Northern areas show particular characters, important for plant growth, of day-length and irradiance which are very different from those of lower latitudes. They have a long photoperiod in the summer lasting for two months or more, with lesser radiant flux densities compared with lower latitudes, differences which are known to have selected biotypes of the same species (e.g. *Oxyria digyna*; Billings & Mooney 1968). Their climates are also obviously very different from those of lower latitudes. Mean annual temperatures are negative, precipitation may be low, with permafrost giving impeded drainage, and the growing season for plants is short. Soil development in these areas will again be very different from further south, often controlled by freeze/thaw action and depth of the active layer above the permafrost table. In large areas of the north, impeded drainage via permafrost has led to the accumulation of peat, with characteristic plant communities of shrub heath tundra; as we shall see, there is little evidence for such tundra communities being characteristic of cold stage vegetation in our area of lower latitude.

Physical processes dependent on freeze/thaw, such as thermal contraction of surficial sediments, cryoturbation and solifluction are related to the subsurface sediment type as well as climate. Their expression in the north under particular conditions does not mean that their presence in cold stages implies the same conditions at lower latitudes.

Thus if we are to search for analogues for East Anglian cold stage conditions we have to find an area where geology may be similar in terms of underlying rock, so at least some comparison is available in terms of physical processes. Such areas are rare in the north. Large regions are alluvium- or peat-covered or have hard basement rocks exposed. Possibilities in north America are parts of the Alaska North Slope, Herschel Island and the western part of Banks Island, where soft sediments form the basis of the regolith and peat is rare or absent. In such areas may be found at least a degree of analogy for vegetation, soils and periglacial processes.

The question of analogues for the flora (and fauna) is much more complex. Whereas physical constants do not change, the Linnean binomial for a plant species gives no indication of the physiological attributes of that species, only of morphological similarity. Such morphological similarity may encompass one or many biotypes and the derived ecotypes, each fitted to its own environment. So the identification of a fossil to a Linnean species cannot link the biotype represented by the fossil to a living biotype of the species. Can there be a physiological analogue of a cold stage fossil in a present northern area when the physical conditions between low and high latitudes are so different? Developments in molecular biology may remedy this situation in the future.

Plants are the foundation of the terrestrial food chain. The correct interpretation of cold stage vegetation and climate has implications for the nature of the co-existent fauna, from insects to large herbivores, as exemplified by the ongoing discussion of the productivity of full-glacial vegetation in Beringia (Laxton *et al.* 1996).

2

Geological setting

The Quaternary Era covers the most recent part of geological time, and is characterised by the climatic changes already mentioned. The Quaternary has been divided into two epochs, Pleistocene and Holocene. The Pleistocene covers the whole of the Quaternary except for the last 10,000 years, which has been separated as the Holocene and covers the time of the present temperate stage. Major periods of distinctive climate during the Quaternary are defined as stages, so that the whole Quaternary can be divided into stages, usually named after type sites or areas. These stages are characterised in north-west Europe in the later Quaternary by alternating temperate and cold conditions. The Holocene includes only the present temperate stage, named in Britain the Flandrian Stage, succeeding the last cold stage of the Pleistocene, the Devensian Stage (Table 2.1).

Since the Pleistocene and the Holocene are very unequal in terms of time, another viewpoint has been to consider the Pleistocene equivalent to the Quaternary, with the present temperate stage being the youngest stage of the Pleistocene. This usage may certainly be found in the literature. However, since the last 10,000 years occupies such an important period, with the development of the present biota after the end of the last cold stage and the period of the rapid technological evolution of humanity, the Pleistocene–Holocene usage receives wider support.

Outline of stratigraphy

An outline of the Quaternary cold stage sequence in Ireland, Britain and The Netherlands is given in Figure 2.1, based on Gibbard *et al.* (1991), Coxon (1993) and Mitchell & Ryan (1997). Only major units are named in this figure; details of subdivisions of the stages are not given.

Table 2.1. *Subdivisions of the Quaternary used for analysis of the cold stage fossil records*

	Cold stage	Subdivision	Abbreviation
		BRITAIN (present temperate stage, Flandrian, Holocene)	
Late Pleistocene	Devensian	Late Devensian	l De
		Middle or Late Devensian	m l De
		Middle Devensian	m De
		Upton Warren Interstadial	m De UW
		Early or Middle Devensian	e m De
		Early Devensian	e De
		Brimpton Interstadial	e De Br
		Chelford Interstadial	e De Ch
		Devensian	De
		(last temperate stage, Ipswichian)	
Middle Pleistocene	Wolstonian	Late Wolstonian	l Wo
		Early Wolstonian	e Wo
		Wolstonian	Wo
		Wolstonian ?	Wo?
		(temperate stage, Hoxnian)	
	Anglian	Late Anglian	l An
		Middle Anglian	m An
		Early Anglian	e An
		(temperate stage, Cromerian)	
	Cromer Complex	Cromer Complex cold stage	Cccs
Early Pleistocene	Beestonian	Late Beestonian	l Be
		Late Beestonian b	l Be b
		Late Beestonian a	l Be a
		Beestonian	Be
	Pre-Pastonian	Pre-Pastonian b/d	PrePa b/d
		Pre-Pastonian d	PrePa d
		Pre-Pastonian c	PrePa c
		Pre-Pastonian b	PrePa b
		Pre-Pastonian a	PrePa a
		Pre-Pastonian	PrePa
	Baventian	Baventian	Ba
		Ludhamian 4c	Lu 4c
		Ludhamian 4b	Lu 4b
	Baventian/Thurnian	Ludhamian 2/4	Lu 2/4

Table 2.1 (*cont.*)

Cold stage	Subdivision	Abbreviation
Thurnian	Ludhamian 2 (Thurnian)	Lu 2(Th)
	Ludhamian 1/2	Lu 1/2
	IRELAND	
	(present temperate stage, Littletonian)	
Midlandian	Middle Midlandian	m Mi
	Early or Middle Midlandian	e m Mi
	Early Midlandian	e Mi
	Early Midlandian	
	Aghnadarragh Interstadial	e Mi Ag
	(last temperate stage)	
Munsterian	Munsterian	Mu
	(temperate stage, Gortian)	
Pre-Gortian	Pre-Gortian	PreG

In north-west Europe the boundary between the Pliocene and Pleistocene has been customarily placed at the base of the Praetiglian Stage of the Netherlands, dated at around 2.3 million years (Gibbard *et al.* 1991). The reason for this placement of the boundary is the Praetiglian is the first stage in the north-west European Cenozoic succession to show the characters of a cold stage. But the Plio/Pleistocene (Tertiary/Quaternary) has been placed by international agreement at 1.7 million years in the Mediterranean Cenozoic succession (Funnell 1995). Since we are dealing with cold stages in north-west Europe, the former usage will be followed.

As seen in Figure 2.1, the most complete Early Pleistocene succession is found in the Netherlands, with the British Early Pleistocene stages correlated broadly with the Praetiglian and Tiglian Stages of the Netherlands. In these Early Pleistocene stages there are complex changes of climate (Gibbard *et al.* 1991), not detailed in Figure 2.1. The correlations given here indicate large gaps in the successions known from the British Isles, with missing cold stages. Our record of cold stage floras is therefore incomplete.

In the Middle Pleistocene, starting at about 700,000 years, the effects of climatic fluctuation appear more acute, and there is a clearer alternation

		IRELAND	BRITAIN	THE NETHERLANDS	Age, years
HOLOCENE		(Littletonian)	(Flandrian)		
PLEISTOCENE	Late	Midlandian	Devensian	Weichselian	10,000
			(Ipswichian)	(Eemian)	125,000
	Middle	Munsterian	Wolstonian	Saalian	
		(Gortian)	(Hoxnian)	(Holsteinian)	
		Pre-Gortian	Anglian	Elsterian	
			(Cromerian (s.s))		
				Cromerian Complex cold events	
	Early			(Bavelian)	700,000
				Menapian	
				(Waalian)	
			Beestonian	Eburonian	1,600,000
			(Pastonian)		
			Pre-Pastonian		
			(Bramertonian)		
			Baventian	Tiglian cold events	
			(Antian)		
			Thurnian		
			(Ludhamian)		
			Pre-Ludhamian	Praetiglian	2,300,000
PLIOCENE				Reuverian	

Figure 2.1. An outline of the sequence of the Quaternary cold stages of Ireland, Britain and The Netherlands. Temperate stages in parentheses.

between temperate stages and cold stages (definitions below), later resulting in the interdigitation of temperate stage sediments with cold stage sediments, often glacial in origin.

In Britain cold stage floras are best preserved in the last cold stage, the Devensian Stage, as might be expected, but they are also known from older cold stages.

Cold stages

Definitions

The term cold stage is used here to include the time intervals characterised by relatively severe climates between the temperate (interglacial) stages of the north-west European Quaternary. Generally they show characters of lithofacies, biofacies and periglacial structures which are nowadays found in areas of cold climate, at higher latitudes or altitudes. These are absent from temperate stages, which are characterised by biofacies demonstrating temperate forest development. This usage is that of Mitchell *et al.* (1973), who discussed the basis for such a subdivision of the British Quaternary.

Subdivision of the Quaternary of the British Isles

The subdivisions used here are based on Coxon (1993), Gibbard *et al.* (1991) and Mitchell *et al.* (1973). Table 2.1 shows the cold stages identified in Britain and Ireland, their subdivision where possible, and the abbreviations for the subdivisions used in the analyses of the floras. Some of the abbreviations apply to the whole stage, others to parts of a stage, since fossil floras are often not clearly related to parts of a stage. The Early Pleistocene includes pre-Cromerian cold stages, but the oldest, the Pre-Ludhamian, recorded in few deep boreholes in East Anglia, is omitted. The Late Pleistocene includes the cold Devensian Stage. In the MACRO and POLLEN record tables, the stratigraphical units are given brief prefixes in alphabetical order from youngest to oldest to facilitate sorting in age order.

Figure 2.2 shows the limits in Britain of the maximum ice extent in the Pleistocene cold stages and the limit of the most extensive and best-known Devensian ice sheet, that of the Late Devensian ice sheet at about 18,000–20,000 years ago. In Ireland the limit of the Midlandian ice sheets is shown; in the earlier cold stage, the Munsterian, most of the present land-mass of Ireland is believed to have been covered by ice apart perhaps from nunataks in mountain areas. Even at times of maximum glaciation in Britain there was in the south a wide periglacial area free of ice, as there was in Ireland in the Midlandian Stage. Areas now offshore and submerged by the post-glacial rise in sea level may have remained ice-free, and north of the ice limits nunataks are likely to have occurred in mountain regions, as discussed by Ballantyne & Harris (1994) and in the Lake District by Lamb & Ballantyne (1998).

The Early (pre-50 ka), Middle, and Late Devensian (post-26 ka) sub-stages used here are those put forward by Mitchell *et al.* (1973), except that

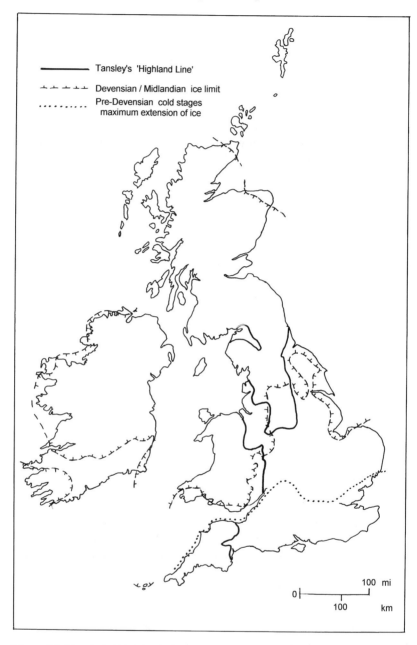

Figure 2.2. Tansley's 'Highland Line' and the limits of the Late Devensian ice sheet in Britain, of the Midlandian ice sheets in Ireland and of the southern limit of ice sheets of earlier cold stages in Britain. An earlier ice sheet in Ireland is thought to have covered most of the country except for nunataks.

records from post *c*. 13 ka, the traditional Devensian late-glacial, are generally excluded. The abundance of Devensian late-glacial sites and records
deserves a separate similar treatment of the rapidly changing vegetation
and flora at the decline of the Devensian, such as that given by Godwin
(1975) and Berglund *et al*. (1996). Interstadials recognised in the Devensian
include the Upton Warren Interstadial Complex of the Middle Devensian
and the Brimpton and Chelford Interstadials of the Early Devensian. The
latter two show the presence of woodland with conifers. The Upton Warren
Interstadial Complex is distinguished by the presence of a more thermophilous Coleopteran fauna (see Chapter 14), but is not so readily identifiable in the recorded flora. This contrast brings to the fore the difficulty of
defining interstadials. The nature of a climatic change will affect biota in
different ways.

The Wolstonian divisions include Late Wolstonian, recorded in sediments underlying temperate Ipswichian Stage sediments, and Early
Wolstonian, overlying temperate Hoxnian Stage sediments. The Anglian
subdivisions include Late Anglian and Early Anglian, with sediments in
analogous positions to the Late and Early Wolstonian sediments. The
Middle Anglian relates to sediments between Anglian tills in East Anglia.
There is a single site (Ardleigh) related to a pre-Anglian Cromerian
Complex cold stage (see Gibbard *et al*. 1991). The older cold stage divisions
are based on West (1980a) and Gibbard *et al*. (1991). The divisions of the
Irish Quaternary are based on Coxon (1993), Mitchell *et al*. (1973) and
Mitchell & Ryan (1997).

The divisions in Table 2.1 reflect the present state of knowledge of subdivision of the terrestrial Quaternary of the British Isles. It is only possible
because of the clear definition of stages in the terrestrial sequence. Stages
additional to those shown in Table 2.1 have been suggested, e.g. by Bowen
(1994). No doubt in future the scheme will be modified as stratigraphical
knowledge accrues. In that case, the sites with fossil flora can no doubt be
allocated to a revised stratigraphy. Any such changes are not likely to affect
the general conclusions reached here on the nature of the cold stage flora.

Dating

Radiocarbon dating has been applied to Late and Middle Devensian sites
with fossil floras. Table 5.5 (p. 67) lists radiocarbon dates associated with
floras. Thermoluminescence dating has been applied to a small number of
Devensian sites with floras, also indicated in Table 5.5. Otherwise, floras are
relatively dated to stages and substages via stratigraphical correlation.

Evidence for cold stage conditions

The evidence derives from lithology of the sediments, from fossils found within the sediments and from structures within the sediments (periglacial structures). The sediments may be inorganic and glacial, fluvioglacial, aeolian, lacustrine or marine in origin, or they may show a much more organic facies, as with lacustrine muds or drift muds. Fossil assemblages may occur *in situ* in all these except the glacial and fluvioglacial sediments, which, however, may contain reworked fossils. The value of such glacigenic sediments is in indicating the extent and timing of glaciation within cold stages rather than giving more exact information about cold stage environmental conditions.

The *in situ* fossil assemblages are a major source of information about cold stage environmental conditions. Also important is the information gained from the presence of periglacial structures, such as thermal contraction cracks, cryoturbation or other structures arising from the former presence of ground ice, indicating perennial or seasonal freezing. Periglacial sediments, such as solifluction diamictons, cover sands and carbonate deposition, may also indicate cold stage conditions. In any interpretation of cold stage conditions, evidence from the fossil assemblages can be strengthened by sedimentary and structural evidence.

Variation of conditions in a cold stage

Within a cold stage there can be no assumption that a uniform climate prevailed. In Britain two conditions have been distinguished within cold stages. Times of severer climate have been termed full-glacial (or pleniglacial), with mainly herbaceous vegetation. In these times, advance of ice sheets to produce glacigenic sediments (e.g. till or boulder clay) occurs. Times of amelioration, but not sufficient amelioration to give a temperate (interglacial) stage, have been termed interstadial. As described above, the interglacial/interstadial distinction was first clearly drawn by Jessen & Milthers (1928) in their study of the Quaternary of Denmark and northwest Germany.

Considering the variability of climatic factors which might produce evidence for full-glacial, interstadial or interglacial conditions, for example precipitation, temperature, or changes of seasonal conditions, the distinctions can only be considered idealistic, in the sense that all transitional conditions between them are possible, indeed likely. But from the present view of subdivision of the Quaternary they can be usefully applied.

A further distinctive character of cold stages appears to be their gradual development and their more abrupt termination, as seen in the marine oxygen isotope curve (Shackleton & Opdyke 1973). This has biological consequences in the terms of the immigration of a cold stage flora, its development and extinction, appearing to give more time for the change from temperate to cold than from the reverse change.

Fossiliferous sediments in periglacial areas

Pollen and macroscopic plant remains are found in a great variety of cold stage sediments in the periglacial area. Many of these are associated with the different facies of braided river systems, others with marine sediments. The variety of sediments leads to a variety of taphonomies for the fossil assemblages, which must be understood to interpret the assemblages effectively. This matter is discussed in the next chapter.

Regional geology

Britain has been divided geologically into two major provinces, one to the north and west underlain by harder and older rocks mainly of Palaeozoic and greater age, and a province to the south and east underlain by softer sedimentary rocks of Mesozoic and Tertiary age. Tansley (1939), in discussing the relation of the present vegetation to this division, called the dividing line 'The Highland Line' (Figure 2.2). He noted the importance of soil development and regional climates in effecting the vegetational contrast between the two provinces, with higher rainfall and poorer soils to the north and west leading to the development of mires and restriction and paucity of woodland.

Similar regionality of soils and climates must obviously occurred in the cold stages in the British Isles. They must have been expressed very differently, for the climatic pattern is likely to have been less meliorated by the Gulf Stream but more influenced by a cold stage climate system related to the Continent. In addition the more rigorous climates must have been associated with cold climate processes which led to frequent soil rejuvenation, with a closer connection between geology and soil than at present. As a result it might be expected that the cold stage flora and vegetation would be more closely related to the nature of the underlying rock.

Unfortunately it is not possible to discern clearly regional variation of the cold stage flora and vegetation because most of the preserved

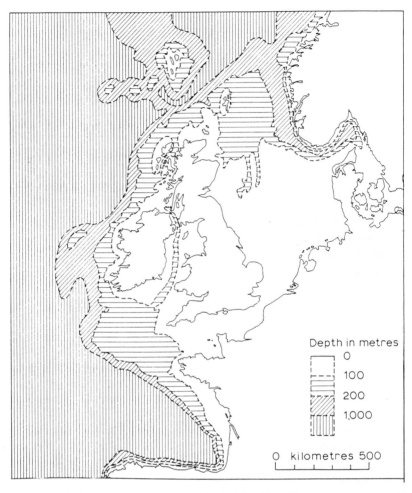

Figure 2.3. Submarine contours around the British Isles. The low sea level of cold
stages gave land bridges to the continent.

Pleistocene fossil assemblages have been found in the south and east province (Figure 5.1, p. 61). Even in this restricted area, though, soil conditions
vary greatly, from sand to loam to clay (Tansley 1939, figure 32). The
heavier soils characterise the widespread boulder clay plateau of East
Anglia, with lighter soils on Chalk and Lower Cretaceous outcrops. The
periglacial structures of stripes and polygons found widely on the Chalk
outcrop demonstrate the significance of cold stage climate processes for soil
development. In addition, the variable deposition of aeolian sediments
such as loess and cover sand during cold stages must have affected soil con-

ditions locally and regionally, as on the Alaska coastal plain (Walker & Everett 1991), and gravelly floodplains provided well-drained areas in river valleys very encouraging to plant propagation.

One reason for the abundance of cold stage sites in the south and east province must be related to the presence of extensive cold stage gravel floodplains even in more minor river systems, which have been heavily exploited for gravel extraction. The gravel floodplains are generally in wide river valleys of low gradient, where variation in fluvial activity has produced changes in sedimentation which have effectively preserved plant remains. In contrast, the conditions in areas of higher relief, harder rocks, steeper gradients and narrower valleys have not been so favourable for the preservation of fossil floras.

The analysis of fossil assemblages which follows therefore only reflects in the main the flora and vegetation of the south and east province. But some similarity obtains between these assemblages and those few recorded in the north and west province, and this encourages a 'baseline' interpretation of the assemblages as a product of a widespread flora and vegetation, even if the detailed variations which must have been present are not yet identifiable.

A final comment on the geological setting must refer to the position of the British Isles on the western edge of the continental shelf (Figure 2.3). Cold stage sea levels were much lower than present sea levels, a result of the eustatic lowering of sea level caused by the build-up of ice sheets in the cold stages. For example, at the time of the maximum extent of the ice in the Late Devensian some 18,000–20,000 years ago, sea level is believed to have been lower than present by 100 m or more. This resulted in land bridges between the British Isles and the continent, with the possibility of concomitant exchanges of biota between the two areas.

3

Sedimentary environments and taphonomy

The interpretation of the fossil record concerns two distinct areas. The first is the geological element of the taphonomy of the assemblages and their sedimentary context. The second is the biological element in terms of the identification of the taxa found fossil and their attributes, such as present habitat and distribution, which will hopefully aid interpretation of past environment and climate. The taphonomy and sedimentary context have to be analysed before we can proceed satisfactorily to the biological elements and the interpretation of environment and climate.

Fossils are sediment and are therefore part of the sedimentary record. As such, they are subject to the processes associated with particle distribution and sorting as sediments form, and the origin, i.e. taphonomy, of a fossil assemblage needs to be considered in any interpretation of the assemblage. This is especially important in cold stage floras, where there is much variety in sediment type, and so taphonomy, often in a single section.

Examples can be given of analyses which require consideration of taphonomy: Bell (1970) showed how variable were the numbers of calyces and pollen grains of *Armeria* in the Middle Devensian samples at Earith. In one sample 33 per cent of *Armeria* pollen was recorded, with 1 per cent in other samples (total pollen and spores). At Somersham (West *et al.* 1999), Devensian pollen samples from bar-tail drift muds with leaf floras and abundant *Salix* leaves contain no *Salix* pollen, while a pool sediment contains much *Salix* pollen accompanied by macroscopic remains. Numbers of a particular species, pollen or macro, are not necessarily indicative of abundance in vegetation. Sorting obviously has important effects in deposition of assemblages.

The source of a fossil assemblage will lie in the deposition of pollen and macroscopic plant remains from the contemporary vegetation and of older fossils received through reworking of older sediments and their redeposition. Since erosion and deposition, especially in a fluviatile regime, are char-

Table 3.1. *Sedimentary situations of cold stage
floras*

1. River (gravel) floodplain		
	Active system	coarse sediments
		bar-top drapes
		bar-tail pools
		pools
	Overbank	floodplain
	Abandoned channel	pool or meander fills
2. Freshwater bodies		
	Lakes and pools	kettle-holes, tunnel valleys
		subsidence, thermokarst
		swales
		solifluction
		proglacial
3. Estuarine/tidal		
	Coastal	shallow, tidal
4. Marine		
	Coastal, offshore	deeper water
5. Glaciomarine		
	Coastal	proglacial

acteristic of the frequent changes wrought by Quaternary climatic and sea level changes, it is not surprising that reworked fossils are not uncommon in cold stage sediments. The separation of contemporary from reworked fossils in assemblages is an important aspect of interpretation of cold stage floras (and faunas).

The taphonomy of pollen and macroscopic remains has to be considered separately. Both represent regional and local vegetation in different ways. Both are sorted in different ways in the course of forming assemblages, and both may be reworked in different ways, making interpretation complex.

The character of a cold stage assemblage depends on such factors as the rapidity of sediment deposition, the grain size of the sediment and the proportion of organic to inorganic particles in the sediment. These factors vary in the many sedimentary situations from which assemblages have been recovered, and which are now described. Table 3.1 lists these sedimentary situations. In our area most are associated with the gravel aggradations of major periglacial rivers, where there is a huge diversity of sedimentary environments (from water bodies to drier gravel, sand and silt) but many other environments are represented. Figure 3.1 shows some of these floodplain situations, also illustrated in Figures 3.2 to 3.6 and Plates 1 and 2.

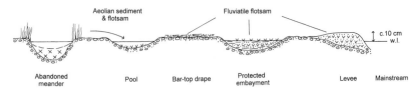

Figure 3.1. Sketch of the gravelly floodplain of a cold stage river, showing the varied origin of fossiliferous sediments. V, organic flotsam; X, organic limnic sediment.

1. Situations associated with a palaeochannel

River floodplains

Knowledge of sedimentation in rivers subject to a periglacial regime is essential to understanding the taphonomy of cold stage floras. Periglacial fluvial regimes have been described by Bryant (1983a,b). The recognition of arctic nival or proglacial regimes, with high stages in the early summer, followed by decreasing discharge, appears to provide a key to the variety of sedimentary environments. The taphonomy of present-day plant remains in Arctic rivers is less well known. Holyoak (1984) has described the accumulation of plant remains in a Spitsbergen river and West *et al.* (1993) have analysed plant remains associated with the floodplain of the River Sagavanirktok on the North Slope of Alaska (Figures 3.2, 3.3). The latter study distinguishes possibilities for the derivation of pollen and macroscopic plant remains in various sedimentary situations, fluviatile and aeolian, indicated in the following discussion. This discussion also refers to Quaternary sites showing sediments formed in particular environments. They contrast with fine-grained blanketing floodplain sediments, alluvial and/or aeolian in origin, which may contain few plant fossils. In such sediments *Chara* encrustations have been recorded, indicating lacustrine conditions in such environments (Boreham & West 1993).

Coarse sediments deposited as sheets, bars or channel or pool fills during high stage may contain reworked fossils or clasts of fossiliferous sediment from bank erosion (e.g. Colney Heath). The reworking may involve a long or very short time span between the time of original deposition and the time of reworking.

Bar-top drapes of sand and silt (Figure 3.4), formed at times of decreasing discharge, may contain pollen and macroscopic plant remains (e.g. Brimpton). The pollen will have been transported with the flow and may be

Figure 3.2. Aerial view of the gravel floodplain of the Sagavanirktok River, adjacent to Franklin Bluffs on the right, flowing north on the Alaska coastal plain. The complexity of channels, bars and vegetation underlies problems of interpretation of the floras preserved in the fine sediments.

Figure 3.3. View of the Sagavanirktok River gravel floodplain near Milepost 346 on the Dalton Highway, Alaska coastal plain, looking east. Tall shrub communities with *Salix* are on the older bars. Moist tundra, looking uniform in the distance, is in the background (West *et al.* 1993).

Figure 3.4. Tail area of a bar on the floodplain of the Sagavanirktok River, Alaska coastal plain. The embayment with the bar-tail silty sand of Figure 3.5 is by the knapsack. Shallow pool fillings and bar-top drapes are also seen.

of more distant origin than the macroscopic plant remains, which are likely to be mostly derived from floodplain vegetation. In such inorganic sediments, there is a possibility that a post-depositional process such as oxidation may affect the composition of the assemblage.

Bar-tail pools (Figures 3.4, 3.5) contain laminated fine sediments with beds of organic flotsam, with rich floras of macroscopic plant remains (e.g. Somersham SAD; Plates 1A, B). The macroscopic plant remains represent local vegetation, swept from the surface of the floodplain in the spring high stage and sedimented in pools at times of later lower discharge. Pollen may be much less abundant, a result of rapid deposition early in summer at times of declining discharge. In the Sagavanirktok study, it was suggested that pollen content varied between that of the inorganic sediment and that of the organic flotsam horizons, with the former containing more coniferous pollen as an 'inorganic fluviatile' component.

In more general terms, **pools** related to active channels may show shallow laminated infillings of sediment determined by frequency of flooding, bringing in fine inorganic sediment and possibly organic flotsam, or by aeolian activity, bringing in silt and sand and organic remains (Plate 1C).

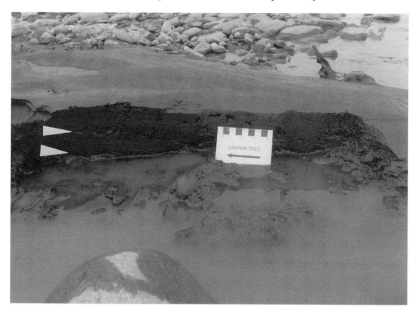

Figure 3.5. Section on floodplain of the Sagavanirktok River showing bar-tail silty sand with two horizons of drift mud rich in plant remains indicated by the arrows (West *et al.* 1993).

Deposition may be rapid and wind may sort organic debris to favour the deposition of coarser material such as leaves (Figure 8.1, p. 138). The sedimentary structures of the inorganic sediments associated with pool sediments give a clue to the taphonomy of the pool assemblages (Plate 2B). Sloughs formed under high energy conditions but then isolated may show leaf floras resting on coarse gravel. Floras with finer and more varied plant remains may occur associated with the lee side of sand ripple structures. Other assemblages can be preserved as skims or drapes at the margins of pools, formed when water levels are lowered and the pool is drained.

Overbank deposition

Sheets of fine sediment occurring in fluviatile sequences have been interpreted as overbank alluvial sediments. The sediments are usually bedded, with interbedded organic drift muds. Reversed grading may be present, showing alternating quietwater conditions interrupted by inflow of coarser material resulting from higher water levels or avulsion (Plate 2A). Examples are sediments of the Weichselian floodplains in the Netherlands described by Ran and Huissteden (1990) and those at Somersham site SAP (West *et al.* 1999).

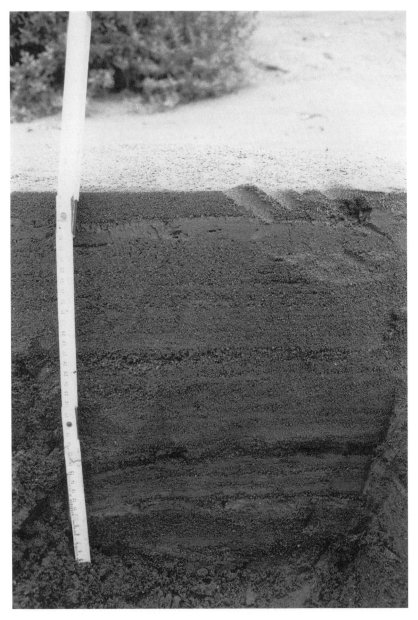

Figure 3.6. Section in a lateral bar of sand and silt, Atigun River area, north of Brooks Range, Alaska. A horizon of drift mud is associated with finer sediment 15 cm from the base (West *et al.* 1993).

Abandoned channels

Water bodies in abandoned channels or pools may contain a much more organic sediment than those associated with active channels, e.g. organic detritus mud, moss peat, sedge peat (e.g. Beetley AA, RR) (Plate 2C). If flooding does occur from time to time, it may be marked by laminations of fine inorganic sediment. Sedimentation appears to be much slower than in the active channel sediments. The pollen content is much richer, and the macroscopic remains content much less varied, than those of the active channel sediments. This results from the more closed nature of the sedimentary environment. There is less of a pollen input from flowing water, with consequently a better representation of local pollen rain. The macro input is dominated by remains from the immediately local vegetation, rather than representing the floodplain sweepings as well.

2. Freshwater bodies

Cold stage floras are well represented in sediments of larger freshwater bodies such as lakes and pools. These may form in proglacial situations, kettle-holes, tunnel valleys, as a result of subsidence of subjacent sediments or through thermokarst or solifluction processes. Their fossil content will depend on the area of the water body and depth of water. Floras are also found in shallow swale conditions where a high water table or impeded drainage has promoted the growth of shallow pool sediments or telmatic organic sediments such as peat.

Lakes and pools

Lakes formed in kettle-holes or tunnel valleys may contain a record of cold stage floras in their basal sediments, formed as ice melted or receded and prior to temperate stage sedimentation (the so-called late-glacial aspect) (e.g. *Hoxne*) or, if the lake has a long life, in a later stage of sedimentation, when cold stage conditions were resumed after temperate stage sediments were formed (so-called early-glacial aspect) (e.g. Marks Tey). If the water is sufficiently shallow in the incipient lake, drift muds with a rich macro flora may be formed (e.g. Hoxne). In deeper water, the pollen flora represents a wider view of the vegetation. In early-glacial conditions, solifluction or ice-rafting may enrich the lake sediments with a good macro flora, adding to the pollen record of regional vegetation from the sediments. But in such a situation there is also the strong possibility of reworking and redeposition of underlying temperate stage sediments as water levels change (e.g. Hoxne, Marks Tey). Under these late-glacial or early-glacial conditions, assessment of the taphonomy of pollen assemblages becomes complex. Pennington (1996) has considered the reworking (or recycling) of pollen in Devensian late-glacial lake sediments via soil instability and the possibility of variation of pollen assemblages as a result of different pollen sources of delivery into a lake, e.g. from ice meltwater or streams draining areas with *Salix* communities.

Subsidence, thermokarst

Lakes or ponds may be formed in smaller depressions formed by collapse or thermokarst processes (e.g. Beetley BD). Again the derivation of plant remains will depend on depth and area of water, governing the type of sediment formed, which itself may be affected by the course of the subsidence or thermokarst development.

Swale pools

Organic telmatic sediments, such as peat, formed in shallow swales in aeolian or fluviatile sediment under conditions of a high water table or impeded drainage are likely to indicate very local conditions of vegetation, with pollen and macroscopic remains very largely derived from the plant communities of the swale (see Ran 1990).

Solifluction

Pools related to solifluction sediments may form on solifluction slopes (e.g. Beetley R). The fossil record in these is valuable in indicating local conditions of vegetation in an area of solifluction. In the Isles of Scilly, Scourse (1991) has described cold stage sediments of ponds associated with active solifluction.

Proglacial

Pools formed in proglacial outwash may contain plant remains (e.g. Tottenhill AA). In such situations, there is the possibility that an assemblage may contain both contemporaneous and reworked plant remains, especially if the ice has overridden older temperate stage fossiliferous sediments. Leaf floras may then give an indication of the local cold stage vegetation, but pollen may be largely reworked and temperate.

3. *Estuarine/tidal*

Sublittoral to intertidal sediments with a flora interpreted as cold stage have been recorded rarely (e.g. Covehithe). Processes of pollen transport are known to be complex in these sediments, so that there are problems of interpretation of pollen spectra in terms of the regional and local vegetation, in particular the separation of the water-borne from the air-borne components of the pollen assemblage (West *et al.* 1980; Brush & Brush 1994; Traverse 1994).

4. Marine

Deeper water assemblages (e.g. Ludham, Sidestrand) suffer from the same problems of interpretation as the inshore assemblages discussed above.

5. Glaciomarine

There is a single record of this environment (Drogheda). Macroscopic plant remains appear to be incorporated as inwash. There is liable to be a high proportion of fossils reworked by glacial action and deposited in the fine sediments formed.

Sedimentary regime

It is useful to take a wider view of controls of the deposition of sediment with organic content in cold stages. In relation to the fluvial environment, snow melt in high Arctic short summers will bring down plant remains over the summer period. There may be no summer low stage, and no deposition of stratified fine sediments as energy stages vary, as may occur in the middle and low Arctic. The latter conditions are more likely to apply to cold stage fluvial conditions at our latitude. But exposures of cold stage fluvial sediments with organic content should be studied with such variability in mind.

Accumulation of organic sediments can be related to drainage impedance by the development of permafrost in sands and gravels. Their occurrence may have no connection with climatic (temperature) amelioration or interstadial conditions, rather the reverse. Where there are periglacial structures associated with such organic beds, these are likely to indicate permafrost degradation, so making permafrost a possible source of the drainage impedance.

Reworking

In considering the taphonomy of an assemblage, the possibility of the presence of a reworked component has to be assessed. Many cold stage processes lead to the reworking of fossils. Major river floodplains in periglacial areas were often occupied in successive cold stages. Moreover, base levels of

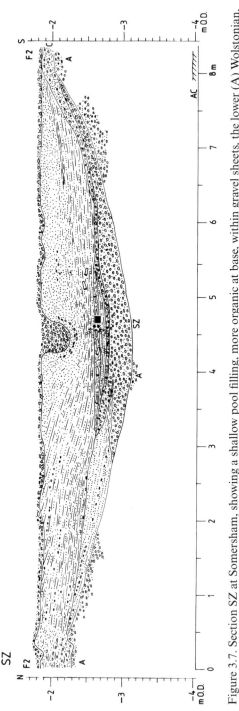

Figure 3.7. Section SZ at Somersham, showing a shallow pool filling, more organic at base, within gravel sheets, the lower (A) Wolstonian, the upper (F2) Devensian. AC, Ampthill Clay. The basin filling contains both temperate stage (e.g. *Taxus* wood, *Corbicula*) and cold stage taxa, a result of the reworking of a temperate flora in Bed C on the right into Devensian sediments. ●, pollen samples; ■, macro sample (West *et al.* 1999).

erosion in these river systems varied following climatic and sea level change. As a result, temperate stage sediments in the same river systems may easily be reworked, resulting in transference of temperate stage fossils to cold stage sediments.

In East Anglia it is not unusual to see sections in fluvial sediments showing a wedge of temperate stage fine organic sediments, gravel above and below, tapering out into what appears to be a single gravel unit (Somersham: West *et al.* 1999; Swanton Morley: Coxon *et al.* 1980). The typical temperate stage mollusc *Corbicula* has been found in cold stage Devensian gravels at the margins of the East Anglian Fenland, and also mixed with cold stage plant remains in cold stage Devensian fine sediments at Somersham (Figure 3.7) (West *et al.* 1999). At Mundesley, on the Norfolk coast, a trunk with roots of the temperate tree *Taxus* was found in cold stage fluvial sediments (West 1980a). In the River Great Ouse valley, in the area downstream from St Ives at Earith, reworked nuts of the temperate taxa *Carpinus* and *Corylus* have been recorded by Bell (1969, 1970). The problem of derivation is considered in more detail by Bell (1969). Such reworked fossils have been found especially near the base of Devensian gravel aggradations in the area, over Jurassic clay. In cutting down to the solid the rivers appear to have incorporated sweepings from the old land surface into the basal Devensian gravels, aggregating cold and temperate stage fossils. Special care is needed in interpreting assemblages from such horizons, for the sediments may contain the sweepings of the land surface and the state of preservation may be no guide.

At Wretton, reworked macroscopic remains and pollen, including many taxa found in nearby older temperate stage terrace sediments, were recorded in a different context. They occurred within the terrace sequence in depressions thought to result from melting ground ice.

An interesting example of bank erosion causing the emplacement of erratic blocks of organic sediment in a fluviatile sequence was seen at Colney Heath, where blocks of organic mud, derived from terrace surface sedimentary environments, were redeposited in a river channel (Godwin 1964). The fossil assemblages gave information on a cold stage flora associated with the terrace surface and dated by radiocarbon to 13,560 BP. In this case the reworking was within the same cold stage.

Solifluction may also be the cause of the introduction of reworked fossils into a sedimentary sequence, perhaps from a temperate stage soil pollen bank. This is likely to be the explanation of the sudden appearance of high *Corylus* pollen frequencies in the sediments deposited at the end of the temperate Ipswichian Stage at Histon Road, Cambridge (Sparks & West 1959).

Though not reworking, but certainly a problem of taphonomy, a special case of contamination of cold stage fossil pollen assemblages by the introduction of recent pollen by tunnelling by bees has been described by Scourse (1991) in the Isles of Scilly, a process which could easily result in errors in interpretation.

Conclusions

Determining the taphonomy of fossil assemblages assists in recognising the possibility and sources of reworking and assessing the regional or local origin of fossil pollen or macroscopic remains. Recognition of reworking in pollen assemblages may require detailed analysis, as that described by Pennington (1996) for Devensian late-glacial lake sediments. Here a study of several sites at different altitudes in the Lake District in northern England revealed the importance of taphonomy in interpreting changes in pollen spectra. In fluviatile sequences, sedimentary context and structures provide the key for considering taphonomy of assemblages.

4

The data tables

The data recorded here are presented in a number of database tables. These are referred to in the text in capitals, e.g. the MACRO table. Fields of the tables are underlined in the text, e.g. taxon, name. The tables are presented as ASCII comma-delimited files (e.g macro.txt) in the CD ROM which accompanies the book. In these files, the fields of the tables are numbered in the way shown in the table descriptions.

Records of macroscopic plant remains, pollen and spores from the cold stage flora sites have been entered into the data base. The compilation started in 1991 and included records published, or provided by authors with permission, to the end of 1995. The basic data of the fossil plant records is given in the MACRO and POLLEN tables. In these tables, each record is related to a sample, a site and a stratigraphical unit. Further descriptive tables give the site location (SITELOC table), the character of the site (SITECHAR table), and dates from sites (CHRONOL table).

Analyses of the information in these tables is assembled to give a list of all taxa identified (TAXA table), and a list of the identified taxa ordered according to *Flora Europaea*, with the number of macro and pollen samples containing each taxon (TAXASORT table). The MCSAMPLE and PLSAMPLE tables list the provenance, taxa numbers and other information for each macro and pollen sample respectively. The representation of the taxa, samples and sites in the stratigraphic units, and therefore in time, is given in the MSAMSTAG, PSAMSTAG and TAXATIME tables.

Further tables are designed to assist in the interpretation of the fossil assemblages. They give details of species identified in terms of biology (PLANT table), habitats (PLANTHAB table) and distribution (PLANTDS1 and PLANTDS2 tables).

The BIBLIO table lists the publications describing the sites and floras, and the other sources for the data used in the compilations.

Table 4.1. *Summary of database tables, fields and records*

Subject	Table	No. of fields	No. of records
Sites	SITELOC	14	94
	SITECHAR	9	195
Chronology	CHRONOL	11	77
Taxa	TAXA	7	1278
	TAXASORT	13	1345
Samples	MCSAMPLE	5	386
	PLSAMPLE	10	750
Taxa records	MACRO	9	6651
	POLLEN	10	13775
Taxa records in time	MSAMSTAG	63	1029
	PSAMSTAG	71	378
	TAXATIME	21	1345
Interpretation of records	PLANT	12	535
	PLANTHAB	11	495
	PLANTDS1	14	488
	PLANTDS2	14	490
Bibliography	BIBLIO	12	104

Table 4.1 summarises the database table titles and the numbers of fields and records in each table. The tables can be usefully grouped by subject as follows:

Sites. Two tables, SITELOC and SITECHAR, describe the site locations and characters.

Chronology. A table, CHRONOL, giving dating results from sites.

Taxa. Two tables, TAXA and TAXASORT, listing the fossil taxa recorded. These are the basis for the systematically-organised list of taxa in Appendix II.

Samples. Two tables, MCSAMPLE and PLSAMPLE, listing macro and pollen samples used.

Taxa records. Two tables, MACRO and POLLEN, listing the macro and pollen records from the sites.

Taxa records in time. Three tables, MSAMSTAG, PSAMSTAG, and TAXA-TIME, showing the occurrence of the macro and pollen records in time.

Interpretation of records. Four tables, PLANT, PLANTHAB, PLANTDS1 and PLANTDS2, with certain information on the biology, habitat and present distribution of taxa recorded fossil.

Bibliography. A table, BIBLIO, giving references to the sources used in compiling the data.

The information in the database tables can now be indicated in detail by describing the structure and fields of each table. The field titles are explained where necessary.

Sites

SITELOC table

The abbreviation and number for each site, and its geographical location and height above sea level. The fields are:

1 Site name: name of site in full.
2 Site: abbreviated name of site.
3 Site no: number of the site on the map, Figure 5.1.
4 VC: Watsonian vice-county or Irish county.
5 OSgr: OS grid letters.
6 OSeast: OS grid easting.
7 OSnorth: OS grid northing.
8 LgWE: W or E longitude.
9 Lg d: degrees longitude.
10 Lg m: minutes longitude.
11 Lt d: degrees latitude N.
12 Lt m: minutes latitude N.
13 Ht: height above sea level.
14 Country.

SITECHAR table

Summary of the character, age, and palaeobotanical information for each site. Abbreviations in Table 4.2. The fields are:

1 Site: abbreviated name of site.
2 Sample: sample title(s).
3 Sedsit: sedimentary situation.
4 Sed: sediments.
5 Age: stage, substage.
6 Abs dat: absolute dating.
7 Rel dat: relative dating, indicating importance of stratigraphy or bio-stratigraphy.
8 No. pl s: number of pollen samples at the site.
9 No. mc s: number of macro samples at the site.

Table 4.2. *Abbreviations for* SITECHAR *and* CHRONOL *tables*

SITECHAR table

Sedsit field

BD	bartop drapes
BT	bar-tail pool fill, stratified, in gravel
C	channel, pool
CG	channel, pool fill in sand, gravel
CP	pool associated with collapse, thermokarst
CS	channel, pool fill on solid
DL	delta
E	estuarine
GM	glaciomarine
L	lake
LK	lake in kettle
M	marine
MF	meander fill in gravel
OB	overbank flood pool (alluvium)
P	pool
PK	pool in kettle
R	reworked
SP	pool associated with solifluction
T	gravel aggradation of terrace
e.g. T P R	reworked erratic from a pool on a gravel terrace
GM L	glaciomarine lake

Sed field

PS	pebbles, stones
S	sand
Z	silt
C	clay
CB	cross-bedded
L	laminated
ST	stratified
CA	calcareous
CM	clay-mud
E	erratic
EG	erratic in gravel
ET	erratic in till
D	detritus mud
DC	detritus mud, coarse
DF	detritus mud, fine/medium
DR	drift mud, stratified with sand, silt
LB	leaf bed
M	marl
OR	organic
P	peat
PB	plant bed, concentrated

Table 4.2 (*cont.*)

PM	moss peat
PW	wood peat
e.g. S DF	sandy fine/medium detritus mud
DF S	muddy sand
S Z	sandy silt
D ET	erratic of detritus mud in till
Z PB L	laminated silt, plant bed

Age field

l	Late
m	Middle
e	Early
st	stadial
i-st	interstadial
De	Devensian
UW	Upton Warren Interstadial
Br	Brimpton Interstadial
Ch	Chelford Interstadial
Wo	Wolstonian
An	Anglian
Be	Beestonian
Cccs	Cromerian complex cold stage
PrePa	Pre-Pastonian
Ba	Baventian
Th	Thurnian
Lu	Ludhamian
Mi	Midlandian
Mu	Munsterian
PreG	Pre-Gortian
e.g. m l De	Middle or Late Devensian
e De Ch	Chelford Interstadial, Early Devensian
e An	Early Anglian

SITECHAR and CHRONOL tables
Abs dat, Rel Dat and Method fields

RC	radiocarbon date
>RC	'greater than' radiocarbon date
(RC)	clearly unreliable radiocarbon date
TL	thermoluminescence date
B	dating by biostratigraphy
S	dating by stratigraphy

Chronology

CHRONOL **table**

Absolute dates for sites. Abbreviations in Table 4.2. The fields are:

1   Site: abbreviated name of site.
2   Sample: title of dated sample.
3   Date publ: publication date of age.
4   Method: dating method.
5   Lab no.
6   RC age BP.
7   TL age.
8   + error.
9   − error.
10 Material: material dated.
11 Comments.

Table 5.5 (p. 67) is an extract from the CHRONOL table giving the dates in order of age.

Taxa

TAXA **table**

Lists taxa recorded fossil, in order in which they were included in the data base; abbreviations in Table 4.3. The fields are:

1   Taxon: the number allocated to the taxon (1–1278).
2   FE fam: *Flora Europaea* family number.
3   FE gen: *Flora Europaea* genus number.
4   FE sp: *Flora Europaea* species number OR the section.
5   FE ssp: *Flora Europaea* subspecies OR section abbreviation.
6   M/P: macro and/or pollen record.
7   Name: *Flora Europaea* name.

TAXASORT **table**

List of fossil taxa in *Flora Europaea* order, plus other records, giving for each taxon the number of samples with pollen and macro records, with annotation (abbreviations in Table 4.4) for the use of the records. Other abbreviations in Table 4.3. The fields are:

Table 4.3. *Abbreviations for* TAXA *and* TAXASORT *tables and the list of taxa (Appendix II)*

agg.	aggregate species
bi	biconvex
cf.	confer; if placed at end of name applies to whole name
CTW	after Clapham,Tutin & Warburg (1962)
FE	*Flora Europaea*
FGB	after Bell (1968, 1970)
g	group
J & T	after Jermy & Tutin (1968); Jermy *et al.* (1982)
ls	large size
m	macroscopic plant remains
nb	non-British
NP	after Polunin (1959)
nfe	non-*Flora Europaea*
p	pollen/spore
s	section or subgenus in FE no. field
sect.	section, subgenus, in name field
s.l.	*sensu lato*
spp.	range of possible species
ssp.	subspecies, variety
subg.	subgenus
t	type
tri	trigonous
x	hybrid
?	possibly, cf.; if placed at end of name applies to whole name
*	identified species
**	identified subspecies
+	identified aggregate species
#	joint species identification

1 Order: numerical order of fossil taxa in *Flora Europaea* (1–1321), plus other taxa recorded (1322–1345).

2 FE fam: *Flora Europaea* family number.

3 FE gen: *Flora Europaea* genus number.

4 FE sp: *Flora Europaea* species number OR the section.

5 FE ssp: *Flora Europaea* subspecies OR section abbreviation or comment.

6 Use: identification category.

7 Name: *Flora Europaea* name.

8 Taxon: taxon number.

9 M/P: macro and/or pollen record.

10 No. macro: number of macro samples with the fossil taxon.

Table 4.4. *Abbreviations for* TAXASORT *table: identification categories in* <u>use</u> *field*

a	species
ay	subspecies
da	species aggregate or *sensu lato*.
b	cf. species or subspecies, if 'a' of that species not present
c	cf. genus and species or subspecies, if 'a' or 'b' of that species not present
ea	species group, if 'a' or 'b' of that species not present
fa	species type, if 'a' or 'b' of that species not present
hb	cf. hybrid
j	postscript to above – identification to more than one taxon
x	postscript to above to indicate probably reworked
m	additional taxon used in analysis tables (habitat, distribution)

11 No. jt macro: number of samples with a record shared with another taxon.

12 No. pollen: number of pollen samples with the fossil taxon.

13 No. jt pollen: number of samples with a record shared with another taxon.

Samples

MCSAMPLE *table*

Lists 386 samples analysed for macro remains. Sites in alphabetical order. Fields as follows:

1 Site: abbreviated name of site.

2 Sample: sample title.

3 No. taxa: number of taxa recorded in sample.

4 No. remains: number of remains in sample.

5 No. frequency: number of taxa recorded by frequency only.

PLSAMPLE *table*

Lists 750 pollen analyses (including 30 duplicates at *Coln* site). Sites in alphabetical order. Abbreviations in Table 4.5. Fields as follows:

1 Site: abbreviated name of site.

2 Sample: sample title.

Table 4.5. *Abbreviations for* PLSAMPLE *and* POLLEN *tables*

Percent base field

tlp	total land pollen
tlp+g	tlp + group concerned
tlp−g (apt)	tlp−group concerned (arboreal pollen total)
tlp−g (cyp)	tlp−group concerned (Cyperaceae)
tlps	total land pollen and spores
tlps+g	tlps + group concerned
tp	total pollen
tps	total pollen and spores
tp−g	tp−group concerned
ttp	total tree pollen
ttp−g	ttp−group concerned

Pres/comm field

b	bad
g	good
LT	low total
p	poor
pr	poor, reworked

Percent category field

Percent in diagram	Category
<1 to 1	1
2	2
3–5	5
6–10	10
11–20	20
21–30	30
31–40	40
41–50	50
51–60	60
61–70	70
71–80	80
81–90	90
91–100	100
101–150	150
151–200	200
201–300	300
301–400	400

N.B. if pollen taxa or other records are extracted, values may be given to the nearest whole figure or 0% if below 0.5%.

3 Percent base: total used for percent calculation.
4 No. taxa: number of taxa recorded in sample.
5 Sample total: total pollen/spores counted.
6 TSP percent: percent tree and shrub pollen in count.
7 Ericales percent: percent Ericales pollen in count.
8 TSP percent base: total used for tree and shrub percent calculation.
9 Pollen assemblage biozone.
10 Pres/comm: preservation, comment.

Taxa records

MACRO table

Lists 6651 records of macroscopic plant remains. Sites in alphabetical order. Abbreviations in Tables 4.6 (plant part) and 4.7 (ident comm and frequency). Fields as follows:

1 Age: stage/substage, with alphabetical prefix.
2 Taxon: taxon number.
3 Site: abbreviated name of site.
4 Sample: sample title.
5 Ident comm: comment on identification.
6 Plant part.
7 Count.
8 Sample total: total remains counted.
9 Frequency: no total.

POLLEN table

Lists 13,775 pollen records. Sites in alphabetical order. Abbreviations in Tables 4.5 (percent category) and 4.7 (ident comm and frequency). Fields as follows:

1 Age: stage/substage, with alphabetical prefix.
2 Taxon: taxon number.
3 Site: abbreviated name of site.
4 Sample: sample title.
5 Ident comm: comment on identification.
6 Frequency: no total.
7 Count.

Table 4.6. *Abbreviations for plant parts in the* MACRO *table*

a	achene	lr	leaf rosette
an	anther	ls	leaf or fruit spine, tip
ann	annulus of sporangium	lsc	leaf scale
b	bud, stem with buds	m	megaspore
bk	bark	ma	massula
bkl	bark, large pieces	mc	mericarp
bsc	bract, budscales	mm	megaspore/microspore
bu	bulbil	my	mycorrhizal root
c	capsule	n	nut, nutlet
ca	calyx	ne	needle
car	caryopsis	o	oospore
cat	catkin	p	pod
ch	charcoal	per	perianth
cl	capsule lid	pet	petiole
co	cone	pl	plug
cpl	carpel	pm	pollen/macro
cs	capsule with seeds	pns	part not specified
csc	cone or catkin scale	pr	prophyll
cv	capsule valve	ps	pod with seeds
d	drupelet	r	receptacle
di	diaphragm (*Equisetum*)	ro	rootstock fragments
e	embryo	s	seed
ep	ephippium	sc	sclerotium
ex	exoderm of fruit	sh	sheath
f	fruit	si	spindle
fl	flower, floret	sp	spikelet
fp	fruiting perianth	st	stem fragment
fsk	fruit stalk	sta	statoblast
fsp	sporangium, lower plant	stl	stem with leaves, shoot
fr	fragment	stp	shoot tip
f/csc	fruit or cone scale	t	tuber
f/s	fruit or seed	tc	tubes of Characeae
fst	fruitstone	te	testa
fv	fruitvalve	th	thallus
g	glume	tr	trunk
i	infructescence	trs	tree stump
ib	involucral bract	tw	twig
in	inflorescence	twl	twig (large species)
l	leaf	u	utricle
lb	leaf base, sheathing	v	valve
le	endocarp lid	w	wood
lf	leaf fragment	ws	wing of seed

Table 4.7. *Abbreviations in* MACRO *and* POLLEN *tables*

(a) For frequency and identification comments

Frequency

a	abundant, common
f	frequent
oc	occasional
r	rare
vr	very rare
p	present
s	seen in scanning (pollen)

Identification comments

a	atypical
ch	charred
d	derived, reworked
d?	questionably derived, reworked
dd	doubtful determination
e	eroded, worn
es	estimated
im	immature
ls	large size
nc	not checked (Lea Valley)
p	part of fine fraction analysed

(b) Prefixes for stratigraphical units, for sorting on age

aa-	l	De			fa-	PrePa	b–d	
ab-	m	l	De		fb-	PrePa	d	
ac-	m	De			fc-	PrePa	c	
ad-	m	De	UW		fd-	PrePa	b	
ae-	e	m	De		fe-	PrePa	a	
af-	e	De						
ag-	e	De	Br		g-	Ba		
ah-	e	De	Ch					
ai-	De				ha-	Lu	4c	
					hb-	Lu	4b	
ba-	l	Wo			hc-	Lu	2/4	
bb-	e	Wo			hd-	Lu2(Th)		
bc-	Wo				he-	Lu	1/2	
bd-	Wo?							
					kb-	m	Mi	
ca-	l	An			kc-	e	m	Mi
cb-	m	An			kd-	e	Mi	Ag
cc-	e	An			ke-	Mi		
cd-	An							
					l-	Mu		
da-	Cccs							
					m-	PreG		
ea-	l	Be						
eb-	l	Be	b					
ec-	l	Be	a					
ed-	Be							

8 Percent.
9 Percent category: percent category from pollen diagram.
10 Percent – taxon: taxon not in pollen sum.

(N.B. if pollen taxa or other records are extracted, values may be given to the nearest whole figure or 0 per cent if below 0.5 per cent).

Taxa records in time

MSAMSTAG table

Lists the number of samples and sites containing macroscopic remains of each taxon in each stage/substage. Fields as follows:

1 Taxon: taxon number.
2 Sample total: total of samples with the taxon.
3 Site total: total of sites with the taxon.
4 1 De samp: number of Late Devensian samples with the taxon.
5 1 De site: number of Late Devensian sites with the taxon.

Likewise a pair of sample and site numbers for earlier stage/substages (abbreviations in Table 2.1), as follows:

6,7. m 1 De	36,37. 1 Be
8,9. m De	38,39. Cccs
10,11. m De UW	40,41. Be
12,13. e mDe	42,43. PrePa d
14,15. e De Br	44,45. PrePa c
16,17. e De Ch	46,47. PrePa b
18,19. e De	48,49. PrePa a
20,21. De	50,51. Lu 2/4
22,23. 1 Wo	52,53. Lu 1/2
24,25. e Wo	54,55. m Mi
26,27. Wo	56,57. e m Mi
28,29. Wo ?	58,59. e Mi Ag
30,31. 1 An	60,61. Mu
32,33. m An	62,63. PreG
34,35. e An	

PSAMSTAG table

Lists the number of samples and sites containing pollen/spores of each taxon in each stage/substage. Fields arranged similarly to MSAMSTAG table:

1 Taxon: taxon number.
2 Sample total: total of samples with the taxon.
3 Site total: total of sites with the taxon.
4 l De samp: number of Late Devensian samples with the taxon.
5 l De site: number of Late Devensian sites with the taxon.

Likewise a pair of sample and site numbers for earlier stage/substages (abbreviations in Table 2.1), as follows:

6,7. m l De	40,41. PrePa b–d
8,9. m De	42,43. PrePa c
10,11. m De UW	44,45. PrePa b
12,13. e m De	46,47. PrePa a
14,15. e De Br	48,49. PrePa
16,17. e De Ch	50,51. Ba
18,19. e De	52,53. Lu 4c
20,21. De	54,55. Lu 4b
22,23. l Wo	56,57. Lu 2/4 ·
24,25. e Wo	58,59. Th
26,27. Wo	60,61. Lu 1/2
28,29. l An	62,63. m Mi
30,31. e An	64,65. e m Mi
32,33. l Be b	66,67. e Mi Ag
34,35. l Be a	68,69. Mu
36,37. Be	70,71. PreG
38,39. PrePa d	

TAXATIME **table**

Lists the occurrence of macros and/or pollen/spores in cold stage stadials and interstadials, temperate stages, present flora, and indicates use of taxa in other tables. Abbreviations in Table 4.8. Fields as follows:

1 Order: numerical order of fossil taxa in *Flora Europaea* (1–1321), plus other taxa recorded (1322–1345).
2 Use: identification category.
3 Name: *Flora Europaea* name.
4 Taxon: taxon number.
5 M/P: macro and/or pollen record.
6 No. macro: number of macro samples with the fossil taxon.
7 No. pollen: number of pollen samples with the fossil taxon.
8 Stadial M: presence of macros in stadia of cold stages.

Table 4.8. *Abbreviations for* TAXATIME *table*

a	absent from PLANTDS1 table: specific identity not known
n	non-British
nr	in temp stage field: not recorded
p	present
x	reworked
?	doubtfully present

9 Stadial P: presence of pollen/spores in stadia of cold stages.

10 UW M: presence of macros in Upton Warren Interstadial Complex.

11 UW P: presence of pollen/spores in Upton Warren Interstadial Complex.

12 Br M: presence of macros in Brimpton Interstadial.

13 Br P: presence of pollen/spores in Brimpton Interstadial.

14 Ch M: presence of macros in Chelford Interstadial.

15 Ch P: presence of pollen/spores in Chelford Interstadial.

16 Temp stage: taxon also recorded in temperate stages.

17 Present: taxon in present British flora.

18 * spp: species identified.

19 x spp: probably reworked taxon.

Interpretation of records

PLANT *table*

Lists properties of taxa which may be valuable for interpretation of fossil records. Abbreviations in Table 4.9. Fields as follows:

1 Taxon: taxon number.

2 Use: identification category.

3 M/P: macro and/or pollen record.

4 Gen/Fam: number of genera in family.

5 Species/Genus: number of species in genus.

6 SSp/Species: number of subspecies in species.

7 Life form: Raunkiaer's life form.

8 Span: life span.

9 Variation: degree of variation.

10 Chrom. no.: diploid chromosome number(s).

11 Repro: reproduction.

12 Poll meth: pollination method.

Table 4.9. *Abbreviations for* PLANT *table*

Use see TAXASORT abbreviations (Table 4.4)
    plus g   genus
         eb   cf. species group, if 'ea' of that group not present

M/P macro or pollen record

Gen/Fam no. of genera in family

Species/Genus no. of species in genus

SSp/Species no. of subspecies in species

Variation
    V   variability remarked
    W   very variable
    P   very polymorphic

Chrom. no.
    2n   numbers
    p   polyploid

Life form   Abbreviations as in Clapham, Tutin & Moore (1989),
             following Raunkiaer

    Phanerophytes
    Ph   woody plants, buds more than 25 cm above soil level
    MM   mega- and mesophanerophytes, from 8 m upwards
    M   microphanerophytes, 2–8 m
    N   nanophanerophytes, 25 cm-2 m

    Chamaephytes
    Ch   woody or herbaceous plants with buds above soil surface
              but below 25 cm
    Chw   woody chamaephytes
    Chh   herbaceous chamaephytes
    Chc   cushion plants

    Hemicryptophytes
    H   herbs (rarely woody plants) with buds at soil level
    Hp   protohemicryptophytes, leafy stems, basal leaves usually smaller
    Hs   semi-rosette hemicryptophytes, leafy stems, lower leaves larger
              than upper, basal internodes shortened
    Hr   rosette hemicryptophytes, leafless flowering stems, basal rosette

    Cryptophytes
    Cr   buds or shoot apices buried in the ground, subdivided as follows:

    Geophytes (geocryptophytes)
    G   herbs with buds below soil surface
    Gb   geophytes with bulbs
    Gr   geophytes with buds on roots

Table 4.9 (*cont.*)

Grh   geophytes with rhizomes
Grt   geophytes with root tubers
Gt   geophytes with stem tubers or corms

Helophytes (limnocryptophytes)
   Hel   marsh plants

Hydrophytes (hydrocryptophytes)
   Hyd   water plants

Therophytes
   Th   pass unfavourable season as seeds

Repro
   – or S seed
   SA apomictic
   V vegetative

Poll method
   A   anemophilous
   C   cleistogamous
   E   entomophilous
   H   hydrophilous
   S   self-pollinating

Span
   1   annual
   2   biennial ( ) rarely
   3   perennial
   S   short-lived

PLANTHAB **table**

Summarises habitats of taxa which may be valuable for interpretation of fossil records. Abbreviations in Table 4.10. Fields as follows:

1 Taxon: taxon number.
2 Use: identification category.
3 Abund: frequency where occurs.
4 M/P: macro and/or pollen record.
5 Veg: vegetation type.
6 Places: type of area where occurs.
7 Water: water requirement.
8 Light: light requirement.
9 Sal: salinity requirement.
10 Soil: soil requirement.
11 Behav: casual, weed, etc.

Table 4.10. *Abbreviations for* PLANTHAB *table*

• ( ) sometimes

Use see TAXASORT abbreviations (Table 4.4)

Abund
  C   common, abundant, frequent
  L   local
  R   rare
  V   very

M/P macro or pollen record

Vegetation
  A   aquatic
  B   bogs
  D   dunes
  Fe   fens, flushes
  Fl   fell, rocks, screes, ledges, crevices,
         barrens
  G   grassland, downs
  H   heath
  Ma   marshes
  Me   meadows
  Mo   moors
  W   woodland, scrub

Light
(not aquatics)
  O   open
  Sh   shady

Water
  Da   damp
  Dr   dry
  Ds   seasonal damp
  We   wet
  Sh   shallow
  Sc   calcareous
  Wa   deeper, lakes
  Wc   ditto, calcareous
  Wo   ditto, oligotrophic

Behaviour
  C   casual
  R   ruderal
  W   weed

Places
  B   bare ground, mud
  Co   coastal
  Cg   cultivated ground
  Ds   disturbed ground
  Fi   fields
  Fl   river/stream sides, banks, lake
         edges
  Gr   grassy places, pastures
  Gv   gravelly places
  H   hedges
  M   mountains
  P   peaty places, muddy
  Ro   road/way sides
  Rp   rocky places, screes, cliffs
  Sa   sandy places
  Sc   scrub, bushy
  Wa   waste places
  Wo   woody places

Salinity
  B   brackish
  M   maritime
  S   saline

Soil
  A   acid, siliceous, base-poor
  C   calcareous, basic, eutrophic
  F   calcifuge, non-calcareous
  H   heavy, clay
           I   inorganic substrate (silt)
  L   light, poor
  M   mull
  O   organic
  S   saline

PLANTDS1 table

Indicates distribution of taxa which may be valuable for interpretation of fossil records. Abbreviations in Table 4.11. Fields as follows:

1 Taxon: taxon number.
2 Use: identification category.
3 Matthews: Matthews' (1937) distribution categories.
4 HG: Godwin (1975): restricted northern range in Scandinavia.
5 EFD S/N: south/north index of Fitter & Peat (1994) (0, most southern; 10, most northern; excl. Ireland).
6 EFD E/W: east/west index of Fitter & Peat (1994) (0, most easterly; 10, most westerly; excl. Ireland).
7 FE: *Flora Europaea* distribution category.
8 Widespread.
9 Arctic.
10 Subarctic.
11 Non-British.
12 Temperate: in temperate stages, including Flandrian or Flandrian archaeological context.

PLANTDS2 table

Relates taxa to the floristic elements of Preston & Hill (1997), giving also numbers of macro and pollen samples containing the taxa. Abbreviations in Table 4.11.

1 Order: order in TAXASORT table.
2 Name: *Flora Europaea* name.
3 Dis 1: Major biome category.
4 Dis 2: Eastern limit category.
5 Disj: distribution disjunct.
6 Cont: continental in Europe.
7 Wid: widely naturalised; present distribution indicated by additional MBC and ELC numbers.
8 Area: a particular distribution noted.
9 FE ssp: *Flora Europaea* subspecies.
10 Use: identification category.
11 Taxon: taxon number.
12 M/P: macro and/or pollen record.
13 No. macro: number of macro samples with the fossil taxon.
14 No. pollen: number of pollen samples with the fossil taxon.

Table 4.11. *Abbreviations for* PLANTDS1 *and* PLANTDS2 *tables*

Use see TAXASORT abbreviations (Table 4.3)

Matthews Matthews' (1937) distribution categories:
M   Mediterranean element
OS   Oceanic Southern element
OW   Oceanic Western European element
ON   Oceanic Northern element
CS   Continental Southern element
CN   Continental Northern element
C   Continental element
NM   Northern-Montane element
AZ   Arctic-Subarctic element
AA   Arctic-Alpine element 'Historical Northern' species
AAB   'Historical Tertiary' species
Al   Alpine

HG   Godwin's (1975) tabulation of taxa with restricted northern range in
       Scandinavia, marked SS

EFDS/N   S/N index of Fitter & Peat (1994) (0, most southern; 10, most
           northern; excl. Ireland)

EFD E/W   E/W index of Fitter & Peat (1994) (0, easterly; 10, westerly)

FE   *Flora Europaea* distribution categories:
A   Widespread in Europe
A1   Throughout most or almost all of Europe
A2   .... but not in extreme north or south
A3   .... but rarer or not in extreme north
A4   .... but rarer in north
A5   .... but rarer or not in extreme south
A6   .... but rarer in south
A7   .... but rarer in Mediterranean
A8   .... but only in mountains in the south
A9   .... but rarer or not in the east
A10   .... but rarer or not in the north-east
A11   .... but rarer or not in the south and east
A12   .... but rarer or not in the south and north-east
A13   .... but rarer or not in south-west and south
A14   .... but rarer or not in the south-east
A15   .... but rarer or not in south-west
A16   .... mainly in mountains
A17   .... but not or rarer in north and west
A18   .... but not or rarer in west
A19   .... but not or rarer in north-west and south

B   Areas of Europe
B1   North
B2   Fennoscandia
B3   West

Table 4.11 (*cont.*)

B4	Central
B5	North, central
B6	North, east, central
B7	North, west, central
B8	North, north-east, central
B9	North-east
B10	North, west
B11	North-west, central
B12	North-west, west
B13	Central, east
B14	Central, west
B15	South, west
B16	South, west, central
B17	South, central
B18	South, central, east
B19	South, south-central, west
B20	South, south-east
B21	South, south-east, west
B22	Mediterranean
B23	South
B24	East
B25	South-west
B26	Balkans
B27	Scotland

C	Arctic, subarctic, montane
C1	Arctic Europe
C2	Arctic, subarctic
C3	Arctic and mountains to south
C4	Arctic, subarctic and mountains to south
C5	Mountains of Norway, Sweden, Iceland
C6	Mountains of Fennoscandia, Scotland
C7	Mountains of central Europe
C8	Mountains of central Europe, Pyrenees, Scotland
C9	Mountains of south-east Europe
C10	Mountains of west, central Europe
C11	Mountains of Europe

D	Montane and wider
D1	North and mountains to south
D2	North, central and mountains to south
D3	North-west and mountains to south
D4	Mountains of central Europe, to north and south
D5	Only in mountains, except in extreme north

Widespread (Clapham, Tutin & Moore 1989; Perring & Walters 1982)	
W	widespread
X	widespread, more local, less abundant
Y	widespread, mainly coastal

Table 4.11 (*cont.*)

Arctic
   A   if so
   AL   alpine

Subarctic
   S   if so

Temperate (Godwin 1975)
   F   in Flandrian temperate stage
   I   in earlier temperate stage
   IC   in Cromer Forest Bed Series
   A   in Flandrian archaeological context

Non-British (Clapham, Tutin & Moore 1989)
   N   if so

Classes of Preston & Hill (1997)
Wid
   W   widely naturalised; present distribution indicated by additional MBC
          and ELC numbers
Area
   Am1 also in N. America
   Am2 in Greenland, not continental N. America
   Am3 very restricted in N. America, more widespread in Europe
   Am4 in west of N. America only in west
   As1   also in central Asia
   As2   also in east Asia
   Bo1   arctic or boreal species, absent from temperate-zone mountains in
           western Eurasia
   Bo2   montane species absent from boreal and arctic in western Eurasia
   Co1   coastal throughout range
   Co2   mainly coastal in British Isles
   Djt   distribution disjunct
   Sst   *sensu stricto*
   Tax   taxonomically difficult
   Tso   in tropics or southern hemisphere or both

BIBLIO table

Lists bibliographical sources of data in alphabetical order of sites. Abbreviations in Table 4.12. Fields as follows:

1 Site: abbreviation name of site.
2 Published: whether published.
3 Form: form of publication.
4 Author(s).

Table 4.12. *Abbreviations for* BIBLIO *table*

Published field	
Y	yes
N	no
Form field	
B	book
J	journal
T	thesis

5 Year.
6 Title.
7 Journal.
8 Volume.
9 Pages.
10 Book.
11 Publisher.
12 Place.

(Text references are listed separately in the text.)

Notes on the database tables

The database tables, presented as ASCII comma-delimited files, are in the accompanying CD ROM disk. Certain extracts from the tables are in the text, e.g. the dates of the CHRONOL table are placed in age order in Table 5.5; the list of identified taxa in *Flora Europaea* order is in Appendix II. The taxa records can be extracted from the MACRO and POLLEN tables in order of age, since there are alphabetical prefixes to the stratigraphical units in the age field of these tables (Table 4.7). The abundance or frequency of taxa in samples can be extracted from the MACRO or POLLEN tables. In the interpretation tables (PLANT, PLANTHAB, PLANTDS1, PLANTDS2), the information, not claimed to be complete or comprehensive, is mainly derived from Clapham, Tutin & Warburg (1962), Clapham, Tutin & Moore (1989) and *Flora Europaea* (Tutin *et al.* 1964–1980). These tables principally concern species; where both species and the related aggregate species are recorded only the species is included.

5

The sites

The sites yielding palaeobotanical data and included in the database are, with very few exceptions, older than *c.* 13 ka, and belong to the Devensian (last) and previous cold stages. If there is a definite uncertainty about the age of a site, it is not included. The list of sites is not certified to be complete. No doubt some sites have been missed. But the great majority of sites with a cold stage flora are included, and the conclusions drawn from the data are deemed to have sufficient validity. Results from new or omitted sites can always be added. Sites presently of doubtful date can be added when their dating is made more secure. If the dating of sites included in the database changes with the acquisition of new data, such changes can be easily accommodated in the data tables. The data from most of the sites included have been published (to end of 1995), but a number of important unpublished sites have also been included by permission of their authors, which permission is gratefully acknowledged. The BIBLIO table shows the source of the data from each site.

The positions of the sites are shown in the map, Figure 5.1. A total of 80 sites is shown on the map. Table 5.1 (extracted from the SITELOC table), lists the sites, gives their number on the map and the abbreviation for the site used in the data tables. Three of the numbered sites cover more than one sampled locality: site 16 covers a number of sites in the Lea Valley (16a to h); site 17 covers four localities in the Colne Valley (Coln, ColnB, Bell, ColnSt); site 79 covers five localities on the Isles of Scilly.

Table 5.2 (extracted from SITECHAR table) shows the number of sites ascribed to particular cold stages or substages. As might be expected, most sites belong to the Devensian (Last) Cold Stage, but the record of cold stage floras goes back to much older parts of the Pleistocene.

Table 5.3 summarises the number of pollen and macro samples and sites in each stratigraphical division.

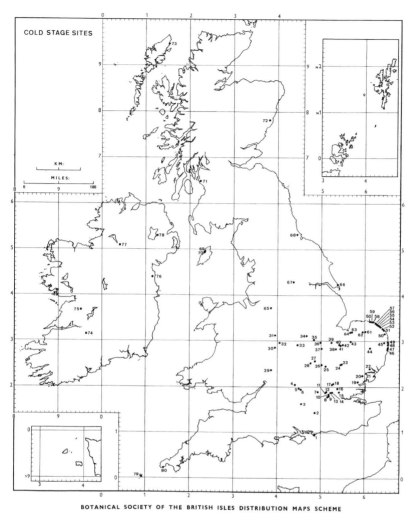

Figure 5.1. Sites with cold stage floras. Numbers/site name key in Table 5.1. Site 16 covers several sites in the Lea Valley. Site 17 covers several sites in the Colney area. Site 79 covers several sites in the Isles of Scilly. Map by permission of the Botanical Society of the British Isles.

Table 5.1 *List of cold stage flora sites included in the data base, with their abbreviations. Site numbers refer to the map, Figure 5.1*

Site no.	Site name	Abbreviation
1	Selsey	Sels
2	Farnham	Farn
3	Brimpton	Brim
4	Stanton Harcourt	Stan
5	Abingdon	Abin
6	Dorchester	Dorc
7	Marlow	Marl
8	Kempton Park	Kemp
9	Twickenham	Twic
10	Isleworth	Isle
11	Colnbrook	Cbrook
12	West Drayton	Dray
13	Kew Bridge	Kewb
14	Ismaili Centre, South Kensington	Isma
15	Ilford	Ilfo
16a	Angel Road, Lea Valley	Ange
16b	Barrowell Green, Lea Valley	Barro
16c	Broxbourne, Lea Valley	Brox
16d	Hedge Lane, Lea Valley	Hedg
16e	Ponder's End, Lea Valley	Pond
16f	Temple Mills, Lea Valley	Temp
16g	Waltham Cross, Lea Valley	Walt
17a	Colney Heath	Coln
17b	Colney Heath B	ColnB
17c	Bell Lane	Bell
17d	Colney Street	ColnSt
18	Hatfield	Hatf
19	Great Totham	Toth
20	Marks Tey	Mark
21	Ardleigh	Ardl
22	Bobbitshole	Bobb
23	Barnwell Station	Barn
24	Sidgwick Avenue	Sidg
25	Sandy	Sand
26	Radwell	Radw
27	Little Houghton	Houg
28	Great Billing	Bill
29	Twyning	Twyn
30	Upton Warren	Upto
31	Four Ashes	Four
32	Nechells	Nech
33	Brandon	Bran
34	Syston	Syst
35	Wing	Wing
36	Thrapston	Thra

Table 5.1 (*cont.*)

Site no.	Site name	Abbreviation
37	Titchmarsh	Titc
38	Hartford	Hart
39	Orton Longueville	Orto
40	Somersham	Some
41	Earith	Eari
42	Block Fen	Bloc
43	Wretton	Wret
44	Hoxne	Hoxn
45	Broome	Broo
46	Easton Bavents	EBav
47	Covehithe	Cove
48	Pakefield	Pake
49	Corton	Cort
50	Ludham	Ludh
51	Happisburgh	Happ
52	Ostend	Oste
53	Bacton	Bact
54	Paston–Mundesley	PaMu
55	Trimingham	Trim
56	Sidestrand	Side
57	Overstrand	Over
58	East Runton	ERun
59	West Runton	WRun
60	Beeston	Bees
61	Swanton Morley	Swan
62	Beetley	Beet
63	Horse Fen, Nar Valley	Hors
64	Tottenhill	Tott
65	Chelford	Chel
66	Dimlington	Diml
67	Oxbow	Oxbo
68	Hutton Henry	Hutt
69	Ballyre, Kirkmichael	Bally
70	Wyllin, Kirkmichael	Wyll
71	Sourlie	Sour
72	Burn of Benholm	Benh
73	Tolsta Head	Tols
74	Baggotstown	Bagg
75	Gort	Gort
76	Tullyallen Quarry, Drogheda	Drog
77	Derryvree	Derr
78	Aghnadarragh	Aghn
79a	Carn Morval, Isles of Scilly	ScCM
79b	Watermill Cove, Isles of Scilly	ScWC
79c	Bread & Cheese Cove, Isles of Scilly	ScBC
79d	Porth Askin, Isles of Scilly	ScPA
79e	Porth Seal, Isles of Scilly	ScPS
80	St Loy	StLo

Table 5.2. *Number of sites in the cold stages and
substages. Certain sites (e.g. Beeston) include more than
one recorded sampling point. See* SITECHAR *table, from
which the numbers are extracted*

Stage, Substage (abbreviations in Table 2.1)	No. of sites	Stage, Substage	No. of sites
l De	16	PrePa b–d	1
m l De	9	PrePa d	4
m De	25	PrePa c	1
m De UW	7	PrePa b	1
e m De	2	PrePa a	7
e De	5	PrePa	1
e De Br	1		
e De Ch	5	Ba	2
		Ba, Lu 4b,c	1
l Wo	5		
e Wo	3	Lu 2/4	1
Wo	4	Th	1
Wo ?	1		
		m Mi	1
l An	8	e m Mi l	
m An	1	e Mi Ag	1
e An	3		
		Mu	1
Cccs	1		
		PreG	2
l Be	1		
Be	1		

Table 5.4 (extracted from SITECHAR table) summarises the frequency of
the sedimentary situations (Table 3.1) of cold stage floras. Most are related
to fluviatile conditions.

A brief survey of the representation of sites in cold stages

BRITAIN

Devensian

The majority of the sites relate to periglacial river gravel floodplains, with a
few lacustrine in origin, some possibly through subsidence. Table 5.5
(extracted from CHRONOL table) shows radiocarbon and TL ages from
Devensian sites with floras (for details of samples, error values, etc., see

Table 5.3. Numbers of analysed samples (pollen and macro) and sites in the stratigraphical subdivisions

Age	Pollen sample	Pollen site	Macro sample	Macro site
l De	139[a]	12	66	13
m l De	11	2	9	7
m De	146	17	92	21
m De UW	19	6	9	4
e m De	14	2	6	1
e De Br	9	1	–	–
e De Ch	45	5	14	4
e De	82	5	55	5
De	19	4	25	5
l Wo	29	5	17	4
e Wo	31	2	17	3
Wo	19	4	9	1
Wo?	–	–	1	1
l An	33	7	19	6
m An	–	–	4	1
e An	6	3	10	3
l Be b	4	1	–	–
l Be a	3	1	–	–
l Be	–	–	3	1
Cccs	–	–	4	1
Be	17	1	4	1
PrePa b–d	1	1	–	–
PrePa d	8	4	1	1
PrePa c	5	1	2	1
PrePa b	6	1	1	1
PrePa a	39	7	4	1
PrePa	1	1	–	–
Ba	17	2	–	–
Lu 4c	3	1	–	–
Lu 4b	4	1	–	–
Lu 2/4	3	1	3	1
Th	11	1	–	–
Lu 1/2	5	1	1	1
m Mi	2	1	1	1
e m Mi	1	1	1	1
e Mi	1	1	2	1
Mu	6	1	3	2
PreG	11	2	3	2
Total	750[a]		386	

[a] This total includes 30 Colney Heath duplicate samples, so real totals are 109 (l De) and 720 (total).

Table 5.4. *Number of sites showing*
particular sedimentary situations in the
cold stages; numbers extracted from
SITECHAR *table*

Sedimentary situation[a]	No. of sites
BD	1
BT	4
C	3
CG	52
CP	1
CS	5
DL	1
E	2
GM	1
L	14
LK	1
M	13
MF	1
OB	3
P	11
PK	1
R	7
SP	7
T	4

[a] Abbreviations as in Table 4.2.

CHRONOL table), from which it is seen there is a good record of floras through the Late and Middle Devensian to the limit of radiocarbon dating.

The Devensian sites do not contain the same kind of vegetational record as has been often found in north-western Europe (e.g. Behre 1989), where deep kettle-holes in Saale/Warthe moraines received sediment though the Last Temperate Stage (Eemian) and parts of the Last Cold Stage (Weichselian) giving great detail of regional vegetation, especially that of the Early Weichselian interstadials.

Wolstonian

Late Wolstonian sites relate to lake, channel or gravel floodplain situations which may later accumulate temperate stage sediments. Early Wolstonian sites relate to similar situations, with certain sites post-dating temperate stage sediments. Wolstonian sites relate to gravel floodplains or proglacial lacustrine situations.

Table 5.5. *Radiocarbon (RC) and TL dates related to floras of the Devensian Cold Stage; data extracted from the* CHRONOL *table, which gives further details of the dates*

Site	Method	RC age BP	TL age
Dray	RC	11230	
Beet	RC	11920	
Bally	RC	12150	
Abin	RC	13260	
Cbrook	RC	13405	
Coln	RC	13560	
Abin	RC	13580	
ColnSt	RC	14320	
Beet	RC	16500	
Diml	RC	18240	
Some	RC	18310	
Diml	RC	18500	
Bloc	RC	18550	
Some	RC	18750	
Barn	RC	19500	
Some	RC	19680	
ScCM	RC	21500	
LeVa	RC	21530	
ScCM	RC	24490	
ScPA	RC	24550	
LeVa	RC	24630	
ScPS	RC	25670	
Beet	RC	26190	
ScWC	RC	26680	
Tols	RC	27333	
Brim	RC	27400	
LeVa	RC	28000	
Some	RC	28020	
Beet	RC	28120	
Bran	RC	28200	
Bill	RC	28230	
StLo	RC	29120	
Wret	RC	29120	
Sour	RC	29290	
Brim	RC	29500	
Sour	RC	30230	
Derr	RC	30500	
Four	RC	30500	
Four	RC	30655	
Bran	RC	30766	
Beet	RC	31750	
Bran	RC	32270	
ScWC	RC	33050	

Table 5.5 (*cont.*)

Site	Method	RC age BP	TL age
Sour	RC	33270	
Sand	RC	34055	
ScPS	RC	34500	
Stan	RC	34730	
Kemp	RC	35230	
Farn	RC	36000	
Four	RC	36340	
Farn	RC	37020	
Syst	RC	37420	
Beet	RC	38000	
Isma	RC	38000	
Four	RC	38500	
Oxbo	RC	38600	
Four	RC	40000	
Upto	RC	41900	
Eari	RC	42140	
Four	RC	42530	
Isle	RC	43140	
Aghn	RC	43560	
Abin	RC	47700	
Marl	RC	>31000	
Upto	RC	>40000	
Benh	RC	>42000	
Four	RC	>43500	
Eari	RC	>45000	
Isma	RC	>45000	
Aghn	RC	>46620	
Aghn	RC	>46850	
Aghn	RC	>47350	
Aghn	RC	>47350	
Aghn	RC	>47950	
Aghn	RC	>48180	
Stan	TL		91 ka
Stan	TL		93 ka

Anglian

Late Anglian sites relate to the time immediately succeeding the retreat or melting of the ice which deposited till correlated to the Anglian Stage. They are succeeded by temperate sediments of the Hoxnian Stage. Sites included in the Early Anglian lie between tills of the Anglian Stage and the older temperate sediments ascribed to the Cromerian Stage. The single Middle Anglian site (Corton) occurs between Anglian tills.

Cromerian Complex cold stage

A single site (Ardleigh) is recorded in this pre-Anglian complex of stratigraphic units.

Beestonian

A single Late Beestonian site (West Runton) records the flora immediately preceding the temperate Cromerian Stage. A number of sites on the Norfolk coast are related to the probably complex Beestonian Stage; they lie between temperate sediments correlated with the Cromerian and Pastonian Stages. There may be more than one cold stage in this interval.

Pre-Pastonian

These sites predate temperate sediments of the Pastonian Stage and are in freshwater or marine environments. A separation into pollen biozones is possible.

Baventian, Thurnian

The sites are in estuarine or marine environments.

IRELAND

Midlandian

Early, Middle and Early/Middle Midlandian sites are related to floodplain channel or delta sediments.

Munsterian

Of the two sites, one is lacustrine and post-dates temperate stage sediments of the Gortian, while the other is part of a glaciomarine sequence.

Pre-Gortian

Two sites record a cold stage flora preceding temperate sediments of the Gortian Stage.

The distribution of the sites

The map, Figure 5.1, shows that the spread of sites is highly uneven, most sites occurring in the south-east quadrant of Britain. Apart from the likelihood that the distribution is biased in favour of areas where observation may have been more easy or concentrated, there are evidently geological factors which control distribution.

Most of the Devensian sites and some of the Wolstonian sites are related to gravel floodplains in lowland areas, indicating that major periglacial rivers with a particular annual discharge regime are most likely to show facies which preserve floras (see Chapter 3). Where rivers are minor and/or relief is greater, these conditions may not prevail. But for these conditions to prevail, gravel and sand sources are also required, and these are more likely to be found in areas with substantial Quaternary glacigenic deposition, as in East Anglia and the west Midlands. An additional bias results from sites which are lacustrine and relate to kettle-holes. Particular conditions of glacial deposition and ice decay then control their distribution. The earlier Quaternary sites are related to the marine sequences of East Anglia, on the west margin of the North Sea basin.

The frequency of sites on gravel floodplains is in contrast with the situation of Middle Pleniglacial (Weichselian) floras of the sand region of the eastern Netherlands. Here sands, silts and peats have been deposited in floodplains by fluvial and aeolian processes, with fine overbank sediments prominent (van Huissteden 1990; Ran & van Huissteden 1990). Any comparisons of the cold stage floras in Britain and on the Continent need to take into account such differences and their ecological and taphonomic consequences.

Comments on particular sites

Aghnadarragh (Aghn): reworking appears to have played some part in the assemblages from this site; Unit 6 is labelled as Early Midlandian interstadial (e Mi Ag), Unit 8 as Early/Middle Midlandian (e m Mi).

Bembridge: a site with *Ranunculus hyperboreus* is not certainly cold stage and has been excluded (Holyoak & Preece 1983).

Bovey Tracey: a site with *Betula nana*, of doubtful age, is considered Late Weichselian by Godwin (1975) and has been excluded.

Colney Heath (Coln): the pollen frequencies incorporated in the database are extracted from Godwin (1964), which excludes Cyperaceae from the pollen sum used. Cyperaceae percentages based on total pollen sum including Cyperaceae were later included from original data. These Cyperaceae percentages are given in the POLLEN table with the sample title suffixed 'a'. Total pollen sum percentages of other taxa can be estimated via knowledge of the Cyperaceae percentage, as this is based on the total pollen sum. Gramineae, *Pinus*, Cruciferae and *Artemisia* percentages used in Figures 8.2, 8.4, 8.5 and 8.6 (pp. 141, 144–6) have been recalculated in this way.

Isles of Scilly: problems of taphonomy and radiocarbon dating at the Scilly sites are considered in detail by Scourse (1985, 1991).

Sites unpublished or partially published at the time of compilation and their authors: Ardleigh (P.L. Gibbard); Broome (in part C.A. Dickson); Drayton (P.L. Gibbard & A.R. Hall); Ismaili Centre, S. Kensington (P.L. Gibbard & A.R. Hall, now published); Sandy (C. Gao & M.E. Pettit, now published); Sourlie (H. Bos & J.H. Dickson), St Loy (J.D. Scourse), Twyning (P.F. Whitehead).

6

Identification of the flora

Fossil records from cold stage floras have accumulated over many years. They clearly need assembling if we are to understand the nature of the flora and the associated ecological and environmental implications. One of the main aims of compiling the database, therefore, was to make as complete a list as possible of all taxa that have been identified, showing also which part of the plant has been identified, such as pollen, leaf, or seed. The question of the reliability of the identifications has to be borne in mind in assessing the results of the analysis of the fossil record; the analyses of the data are subject to the constraints of identification. For example, the identification of *Ranunculus nemorosus* has been questioned, *R. repens* being a preferred identification (Godwin 1975, p. 124). But the records are there and have to be considered. It is not thought that the flavour of the whole is greatly affected by questionable identification. Appendix I gives a list of works used for the identification of macroscopic remains.

The list of identified taxa

The complete list of identified taxa is given in Appendix II. To me, it was unexpectedly long. The table lists records presence of a taxon in a cold stage flora, as macroscopic remains (m), or as pollen or spores (p). There is no indication of abundance or distribution in time in this table. All taxa of higher plants are recorded, in the form identified by the originators of the records, regardless of such problems as reworking and long-distance trans-port. Lower plants are not included, except for a few taxa (see below). A similar compilation of bryophyte records has been made by Dickson (1973), with associated interpretation of ecological and environmental implications.

72

The list in Appendix II is divided into two major columns. On the left are taxa identified to a level higher than the species level. On the right are specific identifications, related to the families, genera or subgenera listed on the left. The specific identifications are marked to show types of specific identification with the four symbols shown at the base of the list of abbreviations in Table 4.3. This table also includes other abbreviations used in Appendix II.

The nomenclature of the higher plant taxa in Appendix II and the TAXA and TAXASORT tables follows *Flora Europaea* except where otherwise stated (e.g. n FE for non-*Flora Europaea*; CTW for Clapham, Tutin & Warburg (1962); NP for Polunin (1959).

The TAXA and TAXASORT tables

Appendix II is complemented by the TAXA and TAXASORT tables. The TAXA table records the taxa in the order they were recorded in the data base, relating them to the *Flora Europaea* classification. The TAXASORT table lists the taxa in *Flora Europaea* order, plus other records, showing the number of analysed samples, macro or pollen/spores, containing the taxon. This gives an immediate, if approximate, indication of the frequency of occurrence of the taxon in cold stage floras, either as macro remains or pollen/spores or both. A comparison of macro and/or pollen occurrence of each taxon is thus possible.

The use field in the TAXASORT table, abbreviations in Table 4.4, gives identification categories to the identified taxa which are used in the analyses of the data tables. Thus all specific identifications can be readily extracted, or all subspecific identifications, and so on.

The MCSAMPLE and PLSAMPLE tables

These tables list the samples, macro (386) and pollen (750), from each site. The number of taxa identified in each sample is given. In the MCSAMPLE table the number of remains in each sample is recorded and the number of taxa recorded by frequency only. In the PLSAMPLE table are given the base for the percent calculations, the pollen/spore total, the percent tree and shrub pollen total, the percent Ericales pollen, the percent base for the tree and shrub percent, any pollen assemblage biozone mentioned and comments on preservation, etc. The tree/shrub and Ericales frequencies are given to summarise their abundance in the samples as a whole.

The MACRO and POLLEN tables

These tables incorporate the fossil macro and pollen/spore records for each site. The age (stage, substage) of the site is given with each record, with any comments made on preservation, etc. The MACRO record includes for each taxon the type of plant part (abbreviations in Table 4.6), number of remains or their frequency and whole sample total; the last is included to give an idea of the frequency of the taxon in the sample. The POLLEN record gives for each taxon the pollen/spore count, percent, percent category (a figure taken from a pollen diagram) or frequency. The abbreviations are shown in Table 4.5. The record and representation of each taxon in time can be extracted from these tables. As will be known to all pollen analysts, there is great diversity in the presentation of pollen analytical results, for example in pollen totals used, in the basis of percent calculations and in the grouping of taxa. The incorporation of pollen data into the database has proved particularly difficult because of this diversity.

The MSAMSTAG, PSAMSTAG and TAXATIME tables

The MSAMSTAG table lists the number of samples and sites containing each macro taxon in each stage/substage, and the PSAMSTAG table gives the same information for each pollen/spore taxon. These tables thus give distribution in time of the taxon concerned. The TAXATIME table gives the occurrence of each taxon, as macro remains or pollen/spores, in cold stage stadials and interstadials, in temperate stages and in the present flora of the British Isles, with an indication of specific identifications and probably reworked taxa. The table thus relates the occurrence of a taxon in a cold stage to its presence or absence in temperate stages or the present flora. Table 4.8 gives the abbreviations used in the TAXATIME table.

The problem of identification

The database records taxa as identified by the authors concerned. The precision of identification will obviously vary, depending on the experience of the identifier, and the availability of well-identified reference material for comparison with the fossils. There may be a bias within reference collections with, for example, a low representation of non-British European species, and there are certainly many families or genera that are found fossil in cold stages which have a complex taxonomy, making identification of even extant species of a genus very difficult or impossible.

The representation of particular taxa can vary greatly. Some, e.g. *Helianthemum*, are recorded as pollen and a variety of macro plant parts, others only as pollen or a plant part. Ease of identification varies greatly, and it would be useful to have a notation for ease of identification of particular taxa. Such a notation was considered for incorporation in the database, but was found to be difficult to compile in the time available. A greater number of satisfactory specific identifications can be made from macro remains than from pollen, but some taxa can be easily identified to a specific level from pollen.

Pollen analysts use a variety of conventions to name pollen taxa, and may name taxa to a group or type. Such categories do not necessarily relate to a particular level of identification in a flora. The nomenclature of the identified pollen/spore taxa in the database is that of the authors concerned. No attempt can be made to standardise a usage for the records, such as that suggested by Birks (1973), although such standardisation would be most helpful.

Important papers on particular cold stage floras have advanced identification standards considerably. Thus the work of C.A. Lambert (Lambert *et al.* 1963) and F.G. Bell (1969, 1970), with detailed studies of macro fossils found in cold stage floras, made a very important contribution to the subject; and in Britain the building-up of the pollen-reference collections at Cambridge by R. Andrew and elsewhere in the 1950s made identification of cold stage pollen taxa much more satisfactory (e.g. Godwin 1964; Andrew 1970). These developments and much detail of cold stage taxa can be found in Godwin (1975), which provides a necessary adjunct to a study of the database. What is needed now to improve identification is more illustrated study of macro remains and pollen and of the extant related genera concerned.

Comments on the flora tables

Indeterminate wood remains and seeds are generally omitted from the TAXA and TAXASORT tables. In these tables all non-*Flora Europaea* taxa are given authorities in the <u>name</u> field. For example, *Ranunculus* sect. *Chrysantha* CTW, related to the *Flora Europaea* systematic number by '61.19.s', or *Arenaria ciliata* agg. B F.G. Bell. *Rubus fruticosus* is given the number '80.9.76.0'.

The highest *Flora Europaea* family number is 199 (Cyperaceae). For certain non-*Flora Europaea* taxa, lower plant taxa and miscellaneous taxa, a 'family' number of 320 or above has been used. The miscellaneous taxa include taxa recorded as reworked, and unknowns, which include a small stem structure recorded in a number of samples and named *Pettitia* in the TAXASORT table.

The name of a taxon is given as it is reported in the original publication. Thus if a taxon is identified as, say, *Abies, Abies* sp. or *Abies* spp., it is so recorded. 'Cf.' is recorded at the end of the taxon name to allow sorting on name.

In the use field of the TAXASORT table identification categories, such as specific identifications, etc., are given (see table 4.4 for abbreviations). These categories are used in the analyses of the flora. The category 'm' labels identified genera which may be used for analysis because of their relative taxonomic or ecological narrowness. Examples are *Botrychium* and *Ophioglossum*.

In the list of macroscopic remains certain taxa at the Late Anglian Nechells site are recorded indistinguishably as macros and pollen. These are marked 'pm' in the plant part field of the MACRO table. In this list two extra taxa are included, since they often occurred in published fossil lists: they are *Daphnia* ephippia and *Cristatella* statoblasts. These are not in the flora list of Appendix II or in the subsequent analyses. *Dryopteris* type sporangia, recorded as macroscopic remains, occur in other fern genera (e.g. *Thelypteris*).

In the list of pollen records, the analyses from Colney Heath (Coln) are duplicated, the first time without Cyperaceae, the second Cyperaceae only.

The problem of reworking of pollen, discussed in Chapter 3, has to be particularly borne in mind in interpreting pollen spectra. The discussion by Scourse (1991) of the taphonomy of the Isles of Scilly pollen spectra shows the problems involved and their possible extent. The resistant grains of *Pinus, Picea* and Compositae are particularly subject to reworking, and the separation of reworked from contemporary grains is difficult. At Somersham (West *et al.* 1999), it was found possible to make a separation based on comparing numbers with the numbers of reworked pre-Quaternary microfossils.

7

The flora

Introduction

The data tables (Chapter 4) provide the detail of the occurrence of particular identified taxa at particular cold stage sites, together with the age and geological context of the sites. The flora itself is listed in Appendix II and the TAXASORT table.

This chapter, based on these tables, provides a brief family-by-family conspectus on the occurrence of taxa found fossil, in *Flora Europaea* order and numbering. For each family, there are accounts of the representation of the family, the genera and species, as macroscopic remains and/or pollen, with mention of particular points regarding taxonomy, taphonomy, habitat and distribution. The number of samples in which a particular taxon has been recorded is given as a general measure of the frequency of the taxon. The full details are in the data tables. A more extensive discussion of certain of the taxa is given by Godwin (1975) in his accounts of the identified fossils.

1. LYCOPODIACEAE

This homosporous family of herbaceous low-growing plants is recorded in very low frequencies throughout the cold stages. Apart from records of the family, three genera are represented: *Huperzia*, *Lycopodium* and *Diphasium*, each with one species. *Huperzia selago* is recorded in 21 samples, all Devensian except for an Early Pleistocene record. *Lycopodium* is recorded in 53 samples through the Devensian and in previous cold stages, with *L. annotinum* in 24 samples in the Devensian and previous cold stages. *Diphasium alpinum* is recorded in eight samples in the Devensian and Wolstonian. The spore frequencies of all the records of this family are usually very low, 2 per cent or below, very rarely more than 5 per cent.

Lycopodiaceae spores have a thick and resistant exospore and are thus good candidates for reworking and redeposition.

2. SELAGINELLACEAE

This heterosporous herbaceous family is recorded throughout the Devensian and previous cold stages, appearing as a consistent member of the cold stage flora.

Megaspores

Megaspores of *S. selaginoides* are recorded in 69 samples from the Devensian and earlier cold stages. Not infrequently they are found in large numbers in some samples (e.g. Aghnadarragh, Beetley, Sourlie), but more usually they occur in lower frequencies.

Microspores

Microspores of *Selaginella* are recorded in 48 samples from the Devensian and earlier cold stages. *S. selaginoides* microspores are recorded in 33 samples, all Devensian and Midlandian. The freqencies ar mainly less than 2 per cent, rarely up to 7 per cent (Wretton, Early Devensian).

3. ISOETACEAE

The family is heterosporous, but only megaspores have been recorded, and those very rarely. They have been listed as single finds or very rare at four sites in four different cold stages, as *Isoetes*, cf. *Isoetes*, *Isoetes lacustris* and *Isoetes* cf. *histrix*, the last in the Middle Devensian Upton Warren Interstadial Complex. Megaspores (and microspores) are more abundant in the Devensian late-glacial and Flandrian (Godwin 1975), a contrast with the rarity in the earlier part of the Devensian and in earlier cold stages.

4. EQUISETACEAE

The genus *Equisetum* is well represented by stem parts (11 samples) and spores (47 samples). There are no specific identifications.

Macroscopic remains

Fragments of stem or nodal structures (sheath, diaphragm) have been found in low numbers, mostly in the Devensian, rarely in earlier cold stages.

Spores

Of the 47 samples containing *Equisetum* spores, most are Devensian or Late Wolstonian, with only three Late Anglian and none of earlier age. The spore frequencies are usually in single figures, the highest being 27 per cent tlp at Block Fen, in contrast to the high values often seen in the Flandrian.

5. OPHIOGLOSSACEAE

Two genera of this homosporous family are recorded, *Ophioglossum* and *Botrychium*. Both have a thick and resistant exospore, and are likely to survive any reworking.

Ophioglossum is recorded in 28 samples from several cold stages. The frequencies are low, usually below 1 or 2 per cent tlp. *Botrychium* usually occurs in similar frequency, but has been found more widely (135 samples), mostly in the Devensian, but also in earlier cold stages. It appears a more regular member of the cold stage flora than *Ophioglossum*, possibly reflecting the more boreal distribution in Europe of *Botrychium* at the present time compared with *Ophioglossum*.

6. OSMUNDACEAE

Both spores and macroscopic remains have been recorded, the latter much more rarely. The provenance of these fossils is problematic.

Macroscopic remains

Two samples from the Norfolk coast Middle/Early Pleistocene contained a small number of annuli of sporangia identified as *Osmunda*. The sediments are marine, and there is the possibility that the annuli are reworked from an earlier temperate stage.

Spores

Very low frequencies of *Osmunda* spores have been recorded from four Devensian floras. *Osmunda* and *O. regalis* spores have been also recorded in very low frequency in a Wolstonian and a Beestonian flora, both freshwater. In the older cold stages, Pre-Pastonian and earlier, low frequencies of these taxa and *O. claytoniana* type have been recorded in 65 samples in marine sediments. The spores of both *O. regalis* and *O. claytoniana* type occur with higher frequency (5–10 per cent tlp) in the underlying Early Pleistocene Ludhamian temperate stage (West 1960), and again there is the possibility that the spores are reworked into later cold stage sediments.

7–22. POLYPODIACEAE s.l.

Records of this group of families are common throughout the cold stages, but the frequencies are usually very low.

Macroscopic remains

Fern sporangia identified to the family have been recorded rarely in the Early Wolstonian. Sporangia identified as *Dryopteris* type have been found rarely in Devensian, Anglian and Pre-Pastonian sites, with higher frequencies in marine sediments. The floatation powers of sporangia may lead to their wide distribution and possible reworking (see West 1980a, p. 9). There is a single record of *Pteridium aquilinum* from Gort.

Spores

Spores identified as Filicales spores have been recorded from 367 samples from all cold stages. The percentage frequencies are usually two or less, rarely rising to between 10 and 20. Frequencies from marine Early Pleistocene sediments are somewhat higher. Spores identified as *Polypodiaceae* or *Polypodium* (102 samples) are recorded in very low frequency in most of the cold stages, with *P. vulgare* recorded at one Devensian site. *Pteridium* spores have been recorded in 19 samples from various cold stages, and *P. aquilinum* from seven Devensian samples. There are a very few records of other fern taxa, which include *Cryptogramma*, *Cystopteris fragilis*, cf. *Polystichum* type, *Dryopteris filix-mas* type and *Gymnocarpium dryopteris*.

Since the spores identified as Filicales or *Polypodium* type have thick exospores, they are likely to survive reworking. Their frequent occurrence with low percentages may indicate considerable reworking throughout cold stages, with cosequentially little certain information from their presence about environmental conditions.

24. SALVINIACEAE

Several megaspores of *Salvinia natans* were found in Devensian sediments at Wretton, but the state of preservation of the macroscopic flora in which they were found suggested that they were redeposited from local temperate stage (Ipswichian) sediments which contained the same flora.

25. AZOLLACEAE

Megaspores, massulae and microspores of *Azolla* have been recorded infrequently from cold stages. Megaspores and massulae have been found in low numbers in Wolstonian and older cold stage sediments (14 samples), sometimes identified as *A. filiculoides*. Microspores have been recorded in four samples from Beestonian and older cold stages. At a number of the sites concerned *Azolla* occurs in local older temperate stage sediments, and there is the possibility of redeposition, particularly in view of the inconsistent and rare occurrence of the taxa. *Azolla filiculoides* occurs abundantly in Hoxnian temperate stage sediments and in earlier temperate stages, but is thought to have become extinct in Europe in the Wolstonian (Godwin 1975).

26. PINACEAE

Macroscopic remains and pollen identified at the family, genus and species level are widely recorded in the cold stages, especially at interstadial sites. Comments on their occurrence are best given under taxon headings. Identification of macroscopic remains to the taxon Coniferae is also included here.

Pinaceae and Coniferae

Remains from seven samples, mainly Devensian, have been identified to these taxa. The finds include wood fragments, charcoal, mycorrhizal roots and a leaf, and occur at very low frequencies. Reworking is always a possibility with such buoyant remains as these.

Abies

Wood of *Abies* has been recorded from one Early Wolstonian sample, charcoal identified as *Abies alba* from two samples at a Middle Devensian site, and there is a single record of *Abies* cf. *alba* from a Pre-Gortian sample. Pollen identified to *Abies* has been recorded much more widely, in 48 samples of Devensian, Early Wolstonian, Beestonian and earlier cold stages. The pollen percentages are usually very low (2% or less), but where the cold stage sediments overlie Late Hoxnian temperate stage sediments with forest pollen assemblages including *Abies* (e.g. Marks Tey), the percentages may be considerably higher. These higher frequencies have been ascribed to reworking of temperate stage sediments during the early part of the subsequent cold stage. The evidence does not indicate that *Abies* was a member of the cold stage flora, but that the wood may be widely distributed by processes of reworking, as may be the pollen, assisted by the massive exine of the body of the pollen grain and its recognisable wings.

Tsuga

Pollen has been recorded in 22 samples, most of Early Pleistocene age. The percentages are very low and the pollen is considered to be reworked from Early Pleistocene temperate stage sediments, which contain higher frequencies of *Tsuga* pollen in the tree pollen assemblages.

Picea

Macroscopic remains of cf. *Picea* have been found very rarely in an Early and a Late Devensian sample. More frequently, *Picea* has been found in Midlandian interstadial sediments, *Picea abies* in Chelford interstadial deposits at Chelford and Brimpton, *Picea abies* ssp. *obovata* in Chelford interstadial deposits at Brimpton, and *Picea abies* cf. *obovata* in Chelford interstadial sediments at Beetley. In these interstadial sediments large numbers of leaves or seeds have been recorded, together with cones, cone scales, wood and trunks. At Beetley over 4700 leaves were extracted. On the basis of macroscopic remains, *Picea abies*, including probably *P. abies* ssp. *obovata*, was a significant member of the Early Devensian interstadial forest.

Picea pollen occurs in very low frequency throughout the cold stages (283 samples). There is slightly increased representation in the Devensian Chelford interstadial, but not to a degree that might be expected from the frequency of macroscopic remains. Under-representation in the pollen rain can be concluded. There is also increased representation in the Early Pleistocene marine sediments. In some samples (e.g. Somersham SAG), higher frequencies have been shown to result from deposition of reworked pollen. The combined macro and pollen evidence points to *Picea* presence in the Early Devensian interstadial and a wider and diluted distribution of pollen through long-distance aerial transport and/or reworking or bias in marine sedimentation in the cold stages.

Pinus

Macroscopic remains of *Pinus* have been recorded from Devensian and Midlandian interstadial sites, usually rarely, except for the interstadial site at Beetley where 552 leaves identified as *Pinus* subg. *Pinus* were counted. Macroscopic remains identified to *Pinus sylvestris* (or cf.), including leaves, cones and seeds, have been more widely recorded, mostly in Devensian and Midlandian interstadial sediments, much more rarely in stadial sediments.

Excluding the few records of pollen of *Pinus* subg. *Haploxylon* and *P. sylvestris* type, *Pinus* pollen is the third most frequently found pollen taxon in cold stage floras (643 samples; Table 8.5, p. 129). The pollen frequencies are highest (40–80 per cent) in interstadial sediments, where the most abundant macroscopic remains are found. Lower but substantial frequencies are found in the Early Pleistocene marine sediments, related perhaps to a larger input of aerial pollen from nearer forest sources and/or reworking and selective transport in marine conditions. Low frequencies are consistently present in stadial sediments, reflecting long-distance aerial transport, and/or reworking. Figure 8.4 (p. 144) shows the number of samples with particular pollen frequencies at different times in the Devensian cold stage.

27. TAXODIACEAE

Pollen of Taxodiaceae has been recorded in low frequency in a single Early Pleistocene sample of marine sediment.

28. CUPRESSACEAE

Pollen of Cupressaceae has been recorded in two Early Pleistocene samples of marine sediment. Macroscopic remains identified to *Juniperus*, cf. *Juniperus* or *Juniperus communis* occur in the Devensian and Pre-Gortian, sometimes in quantity. Pollen referred to *Juniperus* has been recorded more widely in 122 samples from cold stages back to the Early Pleistocene. The pollen frequencies vary. Low frequencies are usual, but frequencies above 10 per cent occur at some sites.

29. TAXACEAE

Macroscopic remains of *Taxus baccata* (wood, seeds) have been found rarely in cold stage sediments. *Taxus* pollen has been recorded in low frequency in two samples. Where the macroscopic remains and pollen are associated unequivocally with a cold stage flora they are interpreted as reworked, as with the tree stump in fluvial sediments of Early Anglian age at Paston–Mundesley.

30. EPHEDRACEAE

Pollen referred to *Ephedra* or *E. distachya* type has been found in very low frequency in only seven samples in four cold stages. *Ephedra* is not a regular component of the cold stage assemblages.

31. SALICACEAE

Of the two genera in this family, *Salix* is well represented in the fossil record of the cold stages, while *Populus* is very poorly represented. Macroscopic remains and pollen of both genera are recorded.

Salix

'The determination of plants in this genus is often a matter of some difficulty; . . . the freedom with which they hybridize . . .' (Clapham, Tutin & Warburg 1962); 'Hybridization plays a very important role in *Salix*.' (Tutin *et al*. 1964–1980). These comments relating to the problem of identification of species of *Salix* apply even more so to the identification of fossil remains. In spite of the problems, a number of species has been identified from fossil leaves, while a very few pollen identifications relate to a specific level. The abundance and consistency of the record shows that *Salix* was clearly an important, if local, member of the shrub and dwarf shrub flora.

Macroscopic remains

These include leaves, twigs, wood, budscales, catkins and capsules. The genus has been identified in over 160 samples, mainly from the Middle and Late Pleistocene. Both prostrate and shrubby species have been identified. The former include *S. herbacea*, the most commonly identified species (42 samples; cf. 6 samples), *S. reticulata* (3 samples; cf. 3), *S. polaris* (6 samples; cf. 5) and *S. repens* (8 samples; cf. 10). Shrubby species include *S. lapponum* (5 samples; cf. 2), *S. viminalis* (7 samples; cf. 13) and recorded rarely *S. myrsinites*, *S. lanata*, *S. phylicifolia* and *S. arbuscula*. Hybrids between some of these species have been rarely recorded.

Pollen

Salix pollen has been identified in a very large number of samples (386) and is present in all cold stages. The genus is entomophilous, and the normal low frequencies encountered may be expected. But there are some sites, stadial and interstadial, where much higher frequencies are reported, e.g. Somersham SAG. Here, *Salix* macro remains were present at the same levels. The higher frequencies are likely to indicate local *Salix* pollen sources. There are a few identifications in the pollen record of *Salix* species, including *S. herbacea* and *S. viminalis*.

Populus

There are few records of *Populus* in the cold stages. Twigs of *Populus* cf. *tremula* occur at one Devensian interstadial site, and capsules and a flower of cf. *Populus* at four stadial sites. Pollen of *Populus* has been rarely recorded.

32. MYRICACEAE

Pollen identified as *Myrica* has been recorded in only two samples, indicating that the shrub was not or rarely a member of the cold stage flora.

33. JUGLANDACEAE

There are single records of *Juglans* and *Carya* pollen, probably reworked. *Pterocarya* has been recorded in eight samples from the Early Pleistocene and Early Wolstonian, probably reworked from the preceding temperate stages.

34. BETULACEAE

The two genera of this family, *Betula* and *Alnus*, are both well represented in the cold stage flora, but in different ways. Pollen of both is found in a large number of samples, but *Betula* macroscopic remains are also abundant while those of *Alnus* are very poorly represented.

Betula

Species of *Betula* are represented macroscopically by wood, catkins, the winged fruits and cone scales. *Betula* (41 samples) has been recorded in all cold stages except the earliest with the marine facies, and *B. pendula* or *pubescens* in cold stages back to Pre-Pastonian a. *B. pendula* has been recorded rarely (4 samples; cf. 4), and *B. pubescens* also rarely (5 samples; cf. 5). Both have been found in stadial and interstadial sediments. *B. nana* is the most commonly recorded of the *Betula* species, back to the Early Pleistocene (54 samples; cf. 15) and sometimes in high frequency (e.g. Thrapston TL2). *Betula* hybrids may make identification of the fruits difficult; cf. *B.* tree species × *B. nana* has been reported once.

Betula pollen is the fourth most common pollen taxon in the cold stage record (582 samples). The frequencies are very variable, sometimes low (stadial, Middle Devensian, Beetley), sometimes much higher, in both stadials and interstadials. The causes of the variation lie not only in vegetation variation but also in the taphonomy of the assemblages. Pollen identified as *B. nana* (15 samples) or *B. nana* type (26 samples) has been recorded in a number of sites back to the Early Pleistocene.

Alnus

Wood, fruits and cones of *Alnus* have been recorded rarely. *Alnus* or cf. *Alnus* has been recorded in eight samples in very low frequency. *A. glutinosa* has been recorded in nine samples, again in very low frequency, the highest being

11 fruits in the Wolstonian site at Tottenhill. The sediment in which this flora was found overlies Hoxnian sediments with *Alnus* and contains many reworked Hoxnian remains, and *Alnus* in this situation is likely to be reworked. Since *Alnus* fruits are sustantial and woody, it is quite possible that the rare remains of *Alnus* in the cold stage record are a result of reworking.

Alnus pollen is one of the most widely recorded of the pollen taxa. The frequencies are very low (usually 1–2 per cent) in the Devensian, even in the interstadials, suggesting that *Alnus* was not a member of the cold stage flora. Higher frequencies are found in the Early Wolstonian at Marks Tey and in the Early Pleistocene, where questions of taphonomy and reworking arise.

35. CORYLACEAE

Three genera of this family have been recorded, *Carpinus*, *Ostrya* type and *Corylus*.

Carpinus

Rare macroscopic remains of *C. betulus* have been found in seven samples, six Devensian and one Early Wolstonian. They are considered to be derived from the preceding temperate stage (where *Carpinus* is present in the forest asemblages), the woody nuts surviving reworking. The easily identifiable pollen grain of *Carpinus* is more widespread (102 samples), and has been found in all the cold stages at frequencies of 1–2 per cent tlp. The nature of the pollen record tends to confirm reworking as the origin of these records.

Ostrya type

There is a single record ($< = 1\%$) from the Early Pleistocene marine sediments, probably reworked.

Corylus

The record is similar to that of *Carpinus*. There are three records of *C. avellana* nuts in the Devensian/Midlandian, and a more widespread *Corylus* pollen record (177 samples) of very low frequencies, usually 2 per cent or less, in cold stages back to the Early Pleistocene (Table 8.7). Reworking is probably the source of the macroscopic remains and pollen.

36. FAGACEAE

Fagus and *Quercus* are both recorded in the cold stages, and are considered to be reworked.

Fagus

There is a single Middle Devensian pollen record.

Quercus

Single budscales of *Quercus* have been recorded in four Devensian samples. As with *Carpinus* and *Corylus*, the pollen is more widespread (129 samples), occurring in very low frequencies back to the Early Pleistocene.

37. ULMACEAE

Very low frequencies of *Ulmus* pollen occur in all the cold stages back to the Early Pleistocene (81 samples). They are considered reworked.

40. URTICACEAE

Achenes of *Urtica dioica* have been widely found (48 samples) in cold stage sediments back to the Early Pleistocene, usually in very low frequency, but in some samples in abundance. The consistency of occurrence of macroscopic remains would suggest that *Urtica dioica* is a member of the cold stage flora. At present it is considered a plant of the Arctic by Polunin (1959) and has a wide tolerance of habitat from fen to disturbed and drier habitats. Pollen identified as *Urtica* has been recorded in far fewer samples (14), at frequencies of 1 per cent tlp or lower. The small size of the grain and the thin exine may make it sensitive to degradation in sediment or under-representation in analyses.

47. POLYGONACEAE

This largely herbaceous family is represented by five genera: *Koenigia*, *Polygonum*, *Bilderdykia*, *Oxyria* and *Rumex*. *Koenigia*, *Bilderdykia* and *Oxyria* are represented by one species each, *Polygonum* and *Rumex* by several. The substantial fossil record of this family in the cold stages must be related to the open and disturbed habitats favoured by many of the species concerned.

Koenigia

Pollen of *K. islandica* is recorded in very low frequency in four samples from the Scottish sites at Benholm amd Sourlie. It is notable that there are no records from further south. The Devensian late-glacial finds mapped by Godwin (1975) show a similar northern distribution.

Polygonum

Macroscopic remains

P. aviculare is the species most abundantly recorded (46 samples), occurring consistently in cold stage floras, large numbers of fruits being found in some samples. *P. viviparum* has been recorded in the Devensian and Wolstonian (30 samples, mostly bulbils). *P. oxyspermum*, *P. hydropiper*, *P. lapathifolium* and *P. amphibium* have been recorded in fewer samples.

Pollen

The genus is eurypalynous, and a number of species have been identified from pollen. The most commonly recorded are *P. bistorta* type (62 samples), *P. bistorta/viviparum* (53) and *P. aviculare* (44). These pollen records parallel in frequency the macro records of the taxa. There are fewer pollen records of *P. maritima*, *P. persicaria*, *P. amphibium* type, *P. bistorta* and *P. viviparum*.

Bilderdykia

There is one macro record of *B. convolvulus* from the Late Wolstonian. The pollen of this taxon or pollen type has been recorded in low frequency from 12 samples of Devensian and Late Wolstonian age.

Oxyria

Fruits of *O. digyna* have been recorded from 11 samples, ranging from the Devensian to the Early Anglian.

Rumex

Macroscopic remains

The genus has been identified in 59 samples in cold stages back to the Cromerian complex cold stage, usually in low numbers. The most commonly identified species is *R. acetosella* (25 samples; agg. 32 samples). *R. acetosa* has been recorded in eight Devensian samples, sometimes in large numbers, and the following species in a few samples each: *R. angiocarpus*, *R. tenuifolius*, *R. conglomeratus*, *R. maritimus* and *R. palustris*.

Pollen

Rumex pollen has been found in 90 samples back to the Early Pleistocene cold stages, usually in low frequencies. Pollen of *R. acetosella* (or type) has been recorded in 37 Devensian samples, and *R. acetosa* (or type) in 37 samples, mainly Devensian but also Early Pleistocene. Pollen of *R. crispus* (or type) has been recorded twice.

48. CHENOPODIACEAE

Several genera of this family are represented in the fossil record: *Chenopodium, Atriplex, Corispermum, Salicornia* and *Suaeda*. As with Polygonaceae, they are herbs characteristic of open and disturbed habitats, with an added preference for maritime situations. The family is stenopalynous, but seed characters assist greatly with identification.

Macroscopic remains

Seeds identified as *Chenopodium, C.* sect. *Pseudoblitum, C. rubrum, C. murale, C. ficifolium* and *C. album* have been recorded in a small number of samples in cold stages back to the Beestonian, usually in very low numbers. *Atriplex* has a similar frequency and age range, with records for *Atriplex, A. patula, A. hastata* and *A. glabriuscula*. Achenes of the non-British *Corispermum*, a genus of continental distribution preferring sandy or stony habitats, have been found in six samples, cf. *Corispermum* in two samples and *C.* cf. *hyssopifolium* in six samples, all of Devensian or Wolstonian age. Seeds of *Suaeda maritima* have been recorded in five samples of Devensian (3) and Early Pleistocene (2) age, the latter in marine sediments. Seeds ascribed to *Salicornia* and cf. *Salicornia* (3 samples in all) have been found in Early Anglian freshwater sediment and Early Pleistocene marine sediments.

Pollen

Pollen of Chenopodiaceae is consistently present (227 samples) in variable but low frequencies throughout the cold stages.

Taken with the evidence for macroscopic remains, the family was clearly an important member of the cold stage flora. The Early Pleistocene presence is probably associated with the marine conditions then prevalent, but the continuation into the Middle and Late Pleistocene can be associated rather with cold stage terrestrial conditions.

55. PORTULACACEAE

Seeds of *Montia* have been found at a small number of sites, mainly Devensian, but also Early Wolstonian and Early Pleistocene. They have been identified as *M. fontana* (6 samples), *M. fontana* agg. (2) and *M. fontana* ssp. *fontana* (4). Pollen identified as *M. fontana* has been recorded in low frequency in the Devensian at Beetley, where seeds were also found.

57. CARYOPHYLLACEAE

This family, the majority herbs, is very well represented in the cold stage flora. *Flora Europaea* describes 37 genera of Caryophyllaceae, of which 14 have been identified in the flora, many with a number of species. Specific identifications are largely based on seeds, few on pollen. The records extend back to the Early Pleistocene.

Macroscopic remains

Apart from finds identified to the family (30 samples), seeds have been identified to 14 genera and 29 species. The most frequent finds have been *Silene vulgaris* (55 samples), *Cerastium arvense* (29), *Arenaria ciliata* (18), *Lychnis alpina* (15), *Stellaria media* (12), *Stellaria palustris* (10), *Minuartia verna* (10) and *Stellaria graminea* (7). Other notable taxa with fewer finds are *Arenaria gothica*, *Minuartia rubella*, *M. stricta*, *M. sedoides*, *Stellaria crassifolia*, *Herniaria glabra*, *Silene furcata*, *Silene acaulis* and *Dianthus deltoides*.

Pollen

Caryophyllaceae is one of the most frequent pollen taxa found in the cold stage pollen assemblages (405 samples). The pollen frequencies are usually low but variable, sometimes rising to 5–10 per cent. A small number of species have been identified, including *Stellaria holostea*, *Scleranthus perennis* and *S. vulgaris*, and a larger number of taxa named as a 'type' or 'group'.

58. NYMPHAEACEAE

Nymphaea and *Nuphar* are both sparsely represented in the flora.

Macroscopic remains

Nymphaea alba has been recorded at one Devensian site, and *Nuphar lutea* at three Devensian sites (9 samples). *Nuphar* has been recorded in the Late Beestonian.

Pollen

The pollen record is more substantial. *Nymphaea* has been recorded in 18 samples (five Devensian sites and a Pre-Gortian site), and *N. alba* at a Devensian site. *Nuphar* pollen has been found in 18 samples (four Devensian sites, one Late Beestonian and one Early Pleistocene).

Although these records are rather few, the presence of the two genera in a variety of sites of different age, combined with clear identification characters of the seeds and pollen, make it likely that they were both members of the cold stage aquatic flora.

60. CERATOPHYLLACEAE

Ceratophyllum has been recorded macroscopically from a Devensian sample. Fruits of *C. demersum* have been recorded from 10 samples of Devensian and Wolstonian age, and a fruit of *C.* cf. *submersum.* from the Devensian. Pollen of *Ceratophyllum* has been found in one Devensian sample. While the fruits of *Ceratophyllum* are substantial and may bear reworking, the pollen grain is thin-walled and insubstantial. The Devensian record for the genus is very sparse, most records being from Wolstonian sites. The presence of *Ceratophyllum* appears to have been present in the Wolstonian, but its presence in the Devensian is perhaps doubtful.

61. RANUNCULACEAE

Four genera of this family have been recorded in the cold stage flora: *Caltha*, *Ranunculus*, *Myosurus* and *Thalictrum*.

Caltha

Seeds of *Caltha palustris* have been found in 16 samples of Devensian and Early Wolstonian age. Pollen of *Caltha* has been recorded more widely (81 samples), in the Late, Middle and Early Devensian, the Late Wolstonian and the Beestonian.

Ranunculus

Many taxa of this genus have been identified, both macroscopic remains and pollen, covering the wide range of open habitats which characterises the genus. The abundant records date back to to the Early Pleistocene.

Macroscopic remains

Five subgenera have been recognised and 15 species. The most frequently recorded are taxa of damp ground, helophytes or aquatics, including *R.* subg. *Batrachium* (189 samples), *R. sceleratus* (58), *R. flammula* (34), *R. hyperboreus* (28) and *R. lingua* (16). Drier ground species have been not so commonly recorded; they include *R.* subg. *Ranunculus* (41 samples), *R. repens* (28) and *R. acris* (14).

Pollen

Identification to species is rarely attained in the genus. The most frequently recorded taxa are Ranunculaceae (166 samples) and *Ranunculus* (160 samples). *R.* subg. *Batrachium* and *R. acris* have been identified in a few samples. Other taxa have been identified as a species group or type.

Myosurus

An achene of *M. minimus* has been recorded in the Early Anglian at Trimingham.

Thalictrum

Achenes have been recorded as follows: *Thalictrum* (7 samples), *T. alpinusm* (28), *T. minus* (11), *T. minus* agg. (8) and *T. flavum* (38). *Thalictrum* pollen has been very widely recorded (315 samples) in low but variable percentages back to the Early Pleistocene. Pollen identified to *T. alpinum* group and *T. flavum* group has been recorded at Sourlie in the Devensian. The genus, represented by three species, is a consistent member of the cold stage flora.

67. PAPAVERACEAE

Seeds of *Papaver* sect. *Scapiflora* (7 samples) and cf. *Papaver* (3 samples) have been recorded in the Devensian, Wolstonian and Cromerian cold stage complex. *P. radicatum* s.l. has been recorded (macro and pollen) at a Devensian site (Sourlie). *P.* sect. *Scapiflora* was evidently a member of the flora in several cold stages.

68. CRUCIFERAE

This large mainly herbaceous family (108 genera in *Flora Europaea*), with many short-lived plants of dry and open habitats, is well represented in the cold stage flora. Many taxa are present as macroscopic remains, with the varied seed characteristics of the family aiding identification. Far fewer taxa are represented by pollen; the family is stenopalynous, with nearly all identifications to family level, and only three to a more detailed level.

Macroscopic remains

Apart from identifications to the family level, remains, mostly seeds, of nine genera and 15 species are recorded. The records extend back to the Early Pleistocene cold stages. The most widely found remains are Cruciferae seeds and fruit valves (in 53 samples), with *Draba* (37), *Draba incana* (37), *Diplotaxis tenuifolia* (41), *Cochlearia* (9), and *Rorippa islandica* (9). Other taxa are found in fewer samples. Seeds of *Draba* and *Diplotaxis tenuifolia* occur abundantly in certain Devensian and Wolstonian samples. *Barbarea,*

Rorippa, *Cardamine*, *Alyssum*, *Erophila* and *Capsella* occur in some quantity in few Devensian, Wolstonian and Anglian samples.

Pollen

Pollen identified to Cruciferae has been recorded in 251 samples, extending back to the Early Pleistocene cold stages (Figure 8.6). Pollen frequencies are mostly below 2 per cent. Higher frequencies (10–33 per cent) are associated with samples also containing macroscopic remains of *Cardamine*, *Draba* and *Diplotaxis*. Evidently there is a low more widespread pollen rain and higher local frequencies associated with floodplain vegetation of both drier habitats (e.g. *Draba*, *Diplotaxis*) and damper areas adjacent to channels (e.g. *Cardamine*). Only three other pollen taxa have been recorded, two generic (*Hornungia*-t, *Sinapis*-t) and cf. *Cardamine pratensis*.

71. DROSERACEAE

Pollen of *Drosera* has been recorded in one Early Pleistocene sample.

72. CRASSULACEAE

Seeds of *Rhodiola rosea* have been found in a Late Devensian sample and in two samples at the Cromer complex cold stage site at Ardleigh. Pollen of cf. *Sedum* has been recorded in three Devensian samples, and pollen ascribed to *Sedum* type in a further Devensian sample.

73. SAXIFRAGACEAE

Saxifraga has an extensive fossil record of macroscopic remains and pollen. *Chysosplenium* has one Devensian record, an identification of *C. alterniflorum* type.

Macroscopic remains

There are records of the genus, a section and a number of species. The identifications are based on leaves and seeds. The genus is notably large and the taxonomy complex. *Saxifraga* is recorded in 12 samples back to the Early Pleistocene. *S*. sect. *Dactyloides* has one record in the Munsterian. Of the species, *S. oppositifolia* (Plate 8A) has the most records (20 samples) back to the Beestonian, with fewer records for *S. hypnoides* (4 samples), *S. hirculus* (3), *S. cespitosa* (2), *S. tridactylites* and *S. rosacea* (1). Some of these species are also represented by several cf. records. Other species are also represented by cf. records and there are four records for *S. hyp-noides/rosacea*.

Pollen

Pollen of the genus has been recorded in very low frequencies at several Devensian sites. *S. opppositifolia* has been recorded in three samples, and pollen has also been identified to cf. species or species groups or types.

74. PARNASSIACEAE

Seeds of *Parnassia palustris* have been recorded in nine samples of Devensian, Wolstonian and Anglian age, with two further Devensian records of cf. *P. palustris*. There is one Devensian pollen record for *Parnassia* and one for *P. palustris*. *P. palustris* appears as a consistent member of the cold stage flora.

79. PLATANACEAE

Two Devensian pollen records of *Platanus* at very low frequency are likely to result from long-distance transport or contamination.

80. ROSACEAE

Eleven genera of this large family are recorded in the flora. Many species have been identified from seeds, particularly in the genus *Potentilla*, with *Rubus chamaemorus* and some genera (*Geum*, *Sorbus*) added to the list by pollen identification.

Macroscopic remains

Many species have been identified, including *Filipendula ulmaria* (17 samples), *Rubus idaeus* (1), *R. fruticosa* (3), *Sanguisorba officinalis* (1), *S. minor* (4), and *Dryas octopetala* (4). *Potentilla* is recorded by the genus (71 samples) and by nine species, of which the most frequent are *P. anserina* (87 samples), *P. palustris* (29), *P. crantzii* (12), *P. erecta* (11) and *P. fruticosa* (6). Two species of *Aphanes* are also recorded.

Pollen

Rosaceae pollen is recorded in 114 samples, of age back to the Early Pleistocene. *Filipendula* is present in 218 samples in low but variable frequencies. *Potentilla* and *Potentilla* type pollen is present in 147 samples, *Rubus chamaemorus* in 7 samples and *Sanguisorba officinalis* in 22.

The species found fossil are mainly herbaceous, reflecting a width of habitat from aquatic (e.g. *P. palustris*) to open conditions such as are now characteristic of waste places, etc. (e.g. *P. anserina*). But there are also shrubs present (e.g. *P. fruticosa*). *P. anserina* is one of the most widespread and abundant cold stage species, large numbers of achenes often being present in the samples. The abundance reflects the common occurrence of open and disturbed ground, especially on river plains.

81. LEGUMINOSAE

This very large family is represented by six genera and a number of species, mainly identified by their seeds and fruits.

Macroscopic remains

Seeds of several genera have been identified (*Vicia, Medicago, Trifolium*). Seeds or fruits of five species are recorded, including *Medicago sativa* ssp. *falcata* (10 samples), *Onobrychis viciifolia* (9) and *Medicago lupulina* (3). *Vicia/Lathyrus* seeds and identifications, some tentative, of species of these genera have also been recorded.

Pollen

Identification of pollen of the family has been made in 58 samples. Several genera have been identified from pollen, including *Vicia* (8 samples) and *Trifolium* (14). Several generic 'types' have also been identified, including *Ulex, Vicia/Lathyrus, Ononis, Lotus* and *Onobrychis*. Specific identifications have always proved difficult, but *Astragalus alpinus* and *Anthyllis vulneraria* have both been identified, as well as specific 'type' identifications (*Lotus uliginosus*) and tentative determinations of species (*Trifolium pratense*).

The consistent representation of Leguminosae in the floras shows that they were an important component of the floodplain vegetation, as they are at present in the Arctic. All the more widespread Arctic genera (Polunin 1959) are represented in some way, except *Lupinus* and *Hedysarum*. No records (macro or pollen) are present earlier than the Late Wolstonian, the Devensian records forming the bulk of the identifications. This may be associated with the paucity of gravel floodplain floras in pre-Devensian times, though these are present (e.g. Broome, Ardleigh), or the history of Leguminosae distribution in the cold stages.

83. GERANIACEAE

There are few records of this family in the cold stages.

Macroscopic remains

There are a small number of Devensian records of seeds from two Devensian sites, identified as cf. Geraniaceae or cf. *Geranium* or *Geranium* spp.

Pollen

Pollen of *Geranium* (11 samples) and *Geranium/Erodium* (1 sample) has been recorded in very low frequency in the Devensian, Late Wolstonian and Pre-Pastonian.

The combination of macros and pollen records in the Devensian, though few, may indicate the presence of the family in the cold stage flora.

86. LINACEAE

The genus *Linum* is well represented by both seeds and pollen in the cold stage flora.

Macroscopic remains

Apart from one record of cf. *Linum*, *Linum* species are abundantly represented, the oldest records being in the Cromer complex cold stage at Ardleigh. The records refer to *L. perenne*, a complex and difficult taxon, with considerable numbers in some samples. The identifications include *L. perenne* (19 samples), *L.* cf. *perenne* (1), *L. perenne* agg. (32), *L.* cf. *perenne* agg. (1) and *L. perenne* ssp. *anglicum*. *L. catharticum* (8 samples) is recorded in several cold stages.

Pollen

Pollen of *Linum* is also recorded in several cold stages. Pollen, in very low frequency, is variously referred to *L. perenne* (1 sample), cf. *L. perenne* (1), *L. perenne* ssp. *anglicum* (3) and *L. austriacum* type (4). Pollen of *L. catharticum* (5 samples) and cf. *L. catharticum* (1) is also recorded.

Linum is a consistently-present taxon in the cold stage flora. *L. perenne*, a very variable species at present and in the Arctic flora, was clearly an important member of the flora.

87. EUPHORBIACEAE

Apart from one record of *Euphorbia* pollen in the Early Pleistocene, the records for this family are based on fruits and seeds. There is one Devensian record for a seed of *Mercurialis perennis* in the Devensian at Thrapston. *Euphorbia* seeds (7 samples) are recorded in the Devensian and Cromer complex cold stage. Seeds of *E. cyparissias* (10 samples) are present in the Devensian and Cromer complex cold stage, and of *E.* cf. *cyparissias* (2 samples) in the Devensian. *E. cyparissias* has thus been found at several sites, and appears to be a regular member of the cold stage flora.

95. ACERACEAE

Pollen of *Acer* has been recorded at very low frequency in four samples, and, as a tree of temperate distribution, may be considered reworked or of long-distance origin.

99. AQUIFOLIACEAE

Ilex pollen has been found in very low frequency in several cold stages (11 samples). As a tree with a southern distribution, and oceanic in the case of *I. aquifolium*, the origin is considered similar to that of *Acer* pollen.

102. BUXACEAE

There are two records of *Buxus* pollen at Beetley in the Early Devensian, one in the Chelford Interstadial. This southern plant is a doubtful member of the cold stage flora. The origin may be similar to that suggested for *Acer*.

103. RHAMNACEAE

Seeds identified as *Frangula alnus* have been found in Early Wolstonian sediments at Hoxne, but they may be reworked from the Hoxnian sediments immediately below, which are known to be involved in reworking. Pollen identified as *Rhamnus* (1 sample) and *Frangula* (3 samples) has been recorded in very low frequency in Devensian and Wolstonian sites. It is probably reworked.

105. TILIACEAE

Tilia pollen, easily recognisable and resistant to destruction, has been recorded from sites back to the Early Pleistocene. Traces of both *Tilia* (18 samples) and *T. cordata* (10 samples) pollen have been found. They are considered reworked from temperate stage sediments.

108. ELAEAGNACEAE

The characteristic leaf scales of *Hippophae rhamnoides* have been found in three samples from two Late Anglian sites and unspecified remains of *H. rhamnoides* in a sample from a further Late Anglian site. Pollen of *H. rhamnoides* has been recorded in one Late Devensian sample, but pollen identified as *Hippophae* has been recorded in many more samples, from a Late Devensian site, from five Late Anglian sites in sometimes high frequencies, fom a Late Beestonian site and from two Pre-Gortian sites. Though low frequencies of *Hippophae* occur inconsistently in cold stages, the highest frequencies (with macro remains) are associated with the end of the Anglian, when *Hippophae* appears to have spread rapidly on the retreat of Anglian ice.

109. GUTTIFERAE

A seed of *Hypericum tetrapterum* has been recorded from the Middle Anglian site at Corton. Pollen of *Hypericum* has been identified in very low frequency in four samples at the Early Devensian site at Wing. The pollen records may suggest that the genus is a member of the cold stage flora. The status of *H. tetrapterum* is uncertain.

110. VIOLACEAE

This family is represented by the large genus *Viola*. The records are mainly of seeds.

Macroscopic remains

Seeds of *Viola*, have been identified in 74 samples in cold stages dating back to the Early Pleistocene. The frequencies are variable, the seeds sometimes abundant. Identification to a species level has been more difficult, perhaps related to the plasticity of the species and the possibilities of hybridisation. There is a variety of taxa named as cf. or two possible species (joint identifications). *V.* subg. *Melanium* has been identified in Devensian and Late Anglian samples (9), *V. palustris* and *V. tricolor* in six samples, and *V. odorata* and *V. lutea* in one sample.

Pollen

There are only two taxa identified, both Late Devensian and in very low frequency in one sample each: cf. *Viola* and cf. *V. palustris*.

112. CISTACEAE

The genus *Helianthemum* is recorded at many sites by both macroscopic remains and pollen, the former allowing specific identification.

Macroscopic remains

Leaves, capsules and seeds identified as *Helianthemum* (7 samples), cf. *Helianthemum* (2), *H. canum* (45) or *H.* cf. *canum* (9) have been recorded at Devensian and Wolstonian sites. *H. canum* is a very variable species, generally now of southern distribution in Europe, and difficult of subdivision into subspecies of distinct geographical range. It was evidently a successful species in the later cold stages.

Pollen

Helianthemum pollen has been identified in 97 samples, dating back to Late Anglian sites. It is usually found in very low frequencies, but in the Late Devensian site at Somersham higher frequencies, up to nearly 30 per cent, were found in association with macroscopic remains of *H. canum*.

115. ELATINACEAE

A single seed of *Elatine* is recorded from an Early Devensian site, and single seeds of *E. hydropiper* and *E. hexandra* from a Middle Anglian site. The seeds are distinctive and rather fragile, and there seems no reason to consider them reworked.

119. LYTHRACEAE

The distinctive pollen of *Lythrum salicaria* has been recorded in very low frequency in two samples from the Early Devensian site at Beetley.

120. TRAPACEAE

A single record of a fragmentary spine of cf. *Trapa* is recorded from a Middle Devensian site (Ismaili Centre).

123. ONAGRACEAE

Macroscopic remains and pollen of *Epilobium* have been recorded, the latter much more commonly than the former.

Macroscopic remains

Seeds of *Epilobium* have been found in Early and Late Wolstonian sediments (5 samples). Seeds of *E. parviflorum* have been recorded in the Late Wolstonian (2 samples), and of *E. alsinifolium* in a Late Devensian sample.

Pollen

Pollen of *Epilobium* has been recorded at sites back to the Early Pleistocene (*Epilobium*, 60 samples; *Epilobium* type, 2 samples). The pollen frequencies are always low. The consistent appearance of the taxon suggests *Epilobium* was a member of the cold stage flora, but little is known of the species represented.

124. HALORAGACEAE

The three native European species of *Myriophyllum* are well represented by both macroscopic remains and pollen.

Myriophyllum

Nutlets identified to the genus are recorded in two Early Pleistocene sites, and pollen in 15 samples dating back to the Early Pleistocene.

M. verticillatum. Nutlets have been recorded in five samples, Devensian, Wolstonian and Anglian in age. The pollen is more widely found (46 samples), sometimes in considerable frequency, again back to the Early Pleistocene.

M. spicatum. Nutlets of this species are more commonly found than those of *M. verticillatum.* They have been recorded in 33 samples, back to the Early Anglian, with a cf. identification in the Early Pleistocene. The pollen has been recorded, usually in low frequencies back to the Early Pleistocene (50 samples).

M. alterniflorum. Nutlets of this species have been recorded in 16 samples, with a cf. identification in a further sample, all in the Devensian. There are records of the pollen in the Devensian and Midlandian (25 samples), sometimes in high frequency.

M. verticillatum and *M. spicatum* were evidently members of the flora in several cold stages, back to the Early Pleistocene, while *M. alterniflorum* has only been found in the last cold stage.

126. HIPPURIDACEAE

Hippuris is well represented in the flora by fruits (nuts), rarely by pollen.

Macroscopic remains

Fruits identified to the genus have been recorded in four Early Devensian samples. *Hippuris vulgaris* fruits have been recorded far more widely in cold stages, back to the Early Pleistocene (97 samples; 1 cf. sample).

Pollen

Hippuris pollen has been identified in two Middle Devensian samples in low frequency.

H. vulgaris, a highly plastic aquatic plant of circumboreal distribution, was evidently a consistent member of the cold stage flora.

127. CORNACEAE

There are a small number of records of fruitstones and pollen of *Cornus*.

Macroscopic remains

C. sanguinea fruitstones have been identified in four samples, and cf. *C. sanguinea* from two samples. At three of the sites concerned the samples are from sediments overlying temperate stage fossiliferous sediments, and since the fruitstones are substantial and woody, and often appear worn, they are considered to be reworked from older temperate stages. There is also a record from a Middle/Late Pleistocene site (Ponders End) of a sample with *C. suecica*.

Pollen

C. suecica pollen was identified at a Middle Devensian (Brimpton) site and *C. mas* type pollen at another Middle Devensian site (Sourlie).

It is doubtful whether *Cornus* species were members of the cold stage flora; possibly *C. suecica* has a claim, but the records are very rare.

128. ARALIACEAE

Hedera pollen has been found very rarely in Devensian and Wolstonian samples (16). It is considered to be reworked.

129. UMBELLIFERAE

Fossils of this large family are common in cold stage floras. Several genera are represented by macroscopic remains and pollen. Specific identifications rely mainly on remains of the fruits.

Macroscopic remains

Fruits identified as Umbelliferae are recorded in 34 samples back to the Early Pleistocene. A number of species have been identified, mostly of plants of aquatic or riparian habitat, including *Hydrocotyle vulgaris* (6 samples), *Berula erecta* (8), *Oenanthe aquatica* (4), *Cicuta virosa* (7) and *Apium inundatum* (3). A smaller number of species of drier ground habitat have been recorded, including *Anthriscus sylvestris* (3 samples) and *Heracleum sphondylium* (3 samples).

Pollen

Umbelliferae pollen has been recorded widely (289 samples), in variable but usually low frequency, back to the Early Pleistocene. The genera *Hydrocotyle*, *Pastinaca* and *Heracleum* have been recorded in a few samples. *Heracleum sphondylium* and *Daucus carota* have been identified in single samples. There are more identifications of taxa named as genus or species types or groups, or as cf. taxa, of which the most common is *Sium* type (13 samples).

132/3. ERICALES

Pollen tetrads identified to this order have been commonly recorded in the cold stages (164 samples) (Figure 8.8). The frequencies are most often below 3 per cent tlp, but are occasionally higher in the younger cold stages (e.g. in the Early Devensian at Wing), much higher frequencies being assso-ciated with marine sediments of the Early Pleistocene.

132. ERICACEAE

Macroscopic remains and pollen of a number of genera and species have been recorded. The family is an important component of shrub tundra in the Arctic at the present time and the fossils are thus of particular interest for the reconstruction of the cold stage flora. The records are therefore summarised in some detail.

Macroscopic remains

Leaves and seeds of Ericaceae (6 samples), cf. Ericaceae (1) and *Erica* (4) have been identified from Early Pleistocene and later cold stages. *Erica* cf. *tetralix* and cf. *Erica tetralix* have been recorded in single samples from the Middle and Late Devensian respectively. *Bruckenthalia spiculifolia* seeds were identified in the Early Devensian at Beetley (5 samples), with two samples within the Chelford Interstadial showing considerable numbers. Cf. *Bruckenthalia* has been recorded in 18 samples from the Early Devensian at Wing. Cf. *Calluna* (1 sample, Early Devensian) and *Calluna vulgaris* (2 samples, Middle Devensian and Middle Anglian) are the rare records of this genus. There is one record of *Rhododendron ponticum* from the Pre-Gortian, and two of *Arctostaphyos uva-ursi* from the Middle/Late Devensian in the Lea Valley.

The genus *Vaccinium* is better represented. Seeds of *Vaccinium* are recorded in the Early and Middle Devensian (9 samples). *V. oxycoccos* has a single record in the Late Devensian, and *V*. cf. *oxycoccos* a single record in the Late Devensian. Cf. *V. oxycoccos* leaves were found in 10 samples from the Early Devensian at Wing. *V. myrtillus* has been recorded in three samples, two Early Devensian Chelford Interstadial and one Wolstonian. There is one Devensian record of a *V*. cf. *myrtillus* seed.

Pollen

Pollen identified to the Ericaceae has been recorded in 47 samples from the last and previous old stages. The frequencies are usually very low. *Bruckenthalia* pollen has been found in 17 samples from Wing (Early Devensian) and Beetley (Early Devensian Chelford Interstadial), a parallel with the macro record. *Calluna* pollen is much more widely represented (176 samples), with *Calluna vulgaris* pollen identified in seven samples; the records stretch back to the Early Pleistocene (Figure 8.9). The pollen frequencies are usually low, but higher during the Early Devensian Chelford Interstadial. *Rhododendron* pollen has been recorded in three Munsterian samples, and cf. *Loiseleuria procumbens* type pollen in a single Middle Devensian sample.

The record of Ericaceous macroscopic remains and pollen is extensive and varied. The question, to be discussed later, is whether this degree of occurrence indicates the regional presence of any communities of shrub tundra type. The record does not appear to be so substantial as to indicate that heath was an important regional component of the vegetation during stadial conditions, except perhaps in the Early Devensian at Wing. The higher frequencies in the Early Devensian Chelford Interstadial are associated with the spread of coniferous forest at that time.

133. EMPETRACEAE

Empetrum is represented by both macroscopic remains and pollen, the latter being more extensive.

Macroscopic remains

Fruitstones identified as *Empetrum* have been recorded in 16 samples from the Early and Middle Devensian and Early/Middle Midlandian. *E. nigrum* has been identified in three samples, in age Middle and Late Devensian and Munsterian. *E. nigrum* agg. has been identified in two Early Devensian Interstadial samples, and *E. nigrum* ssp. *nigrum* in a Middle Devensian sample.

Pollen

Pollen of *Empetrum* (169 samples) (Figure 8.10) and *E. nigrum* type (8 samples) is found throughout the cold stage record, in very low frequencies (including the Early Devensian Chelford Interstadial), but with occasional higher frequencies in the Early Pleistocene marine sediments.

The record for *Empetrum* suggests it was a consistent but minor member of the cold stage flora.

135. PRIMULACEAE

Apart from one Middle Devensian pollen record of *Lysimachia vulgaris* type, all the records for this family are based on seed identification. Four genera are represented.

Primula

The genus is identified in three Devensian samples, with a cf. identification in a further Devensian sample. *Primula* subg. or sect. *Aleuritia* is identified in four Devensian samples and one Wolstonian sample. *P. elatior/veris* is identified in two Devensian samples, *P.* cf. *farinosa* in six Devensian samples and *P. scotica* in a single Devensian sample. Species of *Primula* were certainly members of the cold stage flora, but the identification of species is not easy.

Androsace

A. septentrionalis, a non-British species, is recorded from the Early Wolstonian and Middle Devensian.

Lysimachia

There are single records for *L. vulgaris* in the Late Wolstonian and *L. thyrsiflora* in the Late Devensian.

Glaux

This the the the best represented of the genera in the family, having records of *G. maritima* in 19 samples from the Devensian, Wolstonian and Cromer complex cold stages, with a cf. identification in a further sample. *G. maritima* is a consistently occurring member of the cold stage flora, and has been classed as an obligate halophyte (Bell 1969).

136. PLUMBAGINACEAE

Apart from one record of a seed of cf. *Limonium vulgare* in the Late Devensian, all the many records of this family are of the genus *Armeria*, with both macroscopic remains and pollen of frequent occurrence in the floras.

Macroscopic remains

Calyces and seeds form the basis for the identification of *Armeria*. The genus *Armeria* has been identified in 24 samples back to the Early Pleistocene, with an additional five cf. identifications. *A. maritima* has been identified in a further 71 samples, again back to the Early Pleistocene, *A. maritima* s.l. in three, *A.* cf. *maritima* in 11 and cf. *A. maritima* in one sample. The frequency of calyces is high in some samples.

Pollen

The pollen records for *Armeria* are also abundant, and are found in the Early Pleistocene, Wolstonian and Devensian cold stages. *Armeria* is recorded in 139 samples, *Armeria* type in three samples and *A. maritima* in

41 samples. Pollen frquencies are usually low, but in some samples reach over 10 per cent tlp.

The taxonomy of *Armeria* is complex, with *A. maritima* a very polymorphic species. Nevertheless, the identification of the fossils to *A. maritima* seems satisfactory, and the species appears as a consistent and important member of the cold stage flora.

139. OLEACEAE

Low frequencies of pollen of *Fraxinus* have been found in four samples. They are considered to be reworked.

140. GENTIANACEAE

The family is represented by few macroscopic and pollen records, belonging to three genera.

Macroscopic remains

Seeds of *Centaurium erythraea* have been identified in one Pre-Pastonian sample, and seeds of *Gentianella* seed have been recorded in one Late Devensian sample.

Pollen

Gentianaceae pollen has been identified in nine Devensian samples and in one Wolstonian sample. *G.* cf. *purpurea* (1 sample), *G. pneumonanthe* (1), *G.* cf. *pneumonanthe* (5), *G. pneumonanthe* type (4), *G. nivalis* (1) and *Lomatogonium rotatum* type have been recorded in the Devensian. All these are in very low frequency. *G. pneumonanthe* is the best represented taxon in the family, perhaps because the preference for damp habitats will have favoured preservation.

141. MENYANTHACEAE

Menyanthes seeds and pollen, both with easily recognisable characteristics, have been identified in a large number of samples.

Macroscopic remains

The seeds of the genus have been recorded in six Devensian samples, and seeds of *M. trifoliata* in 47 samples back to the Early Pleistocene.

Pollen

Pollen identified as *Menyanthes* has been recorded in 41 samples, again back to the Early Pleistocene, with *M. trifoliata* recognised in six Devensian samples. The pollen frequencies are usually very low.

M. trifoliata appears to have been a regular member of the cold stage aquatic flora.

144. RUBIACEAE

While macroscopic remains tentatively identified to *Galium* have only been recorded in two samples at one Late Devensian site, pollen of Rubiaceae, a stenopalynous family, is one of the most frequent pollen taxa in cold stage pollen spectra (304 samples). *Galium* has been identified in a further 20 samples. The pollen frequencies are usually low but can rise to over 30 per cent tlp (Isles of Scilly).

145. POLEMONIACEAE

Pollen of *Polemonium* has been recorded in very low frequency in 23 Devensian samples and one Early Pleistocene sample. Pollen identified as cf. *Polemonium* and *P. coeruleum* has been been recorded in single Devensian samples.

146. CONVOLVULACEAE

Seeds identified as cf. *Cuscuta* have been recorded in a single Devensian sample. Pollen grains identified as *Convolvulus* and *Convolvulus arvensis* have been recorded in single Devensian samples in very low frequency.

148. BORAGINACEAE

Nutlets identified as *Myosotis* have been recorded in two Middle Devensian samples. The pollen record is nearly as poor, and is confined to the Devensian. There are very low frequencies for Boraginaceae (3 samples), *Lithospermum* (2), *Symphytum officinale* (2) and cf. *Myosotis* (1).

150. CALLITRICHACEAE

This family is only recorded by macroscopic remains. The fruits have been found in cold stages back to the Early Pleistocene. There are records for

Callitriche (16 samples), cf. *Callitriche* (2 samples), *C. hermaphroditica* (3), *C.* cf. *stagnalis* (5), *C. obtusangula* (1), *C. platycarpa* (2), *C.* cf. *platycarpa* (1) and *C.* cf. *palustris* (1). This is a taxonomically difficult genus, with much plasticity of form, but it has clearly been a consistent member of the cold stage flora.

151. LABIATAE

This large family is represented by macroscopic remains and pollen. Specific identification relies on nutlets, the pollen identifications being restricted to cf. identifications or to pollen types, each of which may cover a number of genera.

Macroscopic remains

A large number of taxa have been identified. The identifications include the family (3 samples), of genera *Galeopsis* (1 sample), *Stachys* (2), *Mentha* (13) and *Nepeta*? (1). Specific identifications include *Ajuga reptans* (3 samples), *Galeopsis tetrahit* agg. (1), *Stachys sylvatica* (1), *S. palustris* (5), *Prunella vulgaris* (3), *Calamintha sylvatica* (1), *Origanum vulgare* (1), *Lycopus europaeus* (25), *Mentha aquatica* (3) and *M. aquatica/arvensis* (15). Many of these taxa must be associated with aquatic or riparian habitats. The identifications stretch back to the Early Pleistocene.

Pollen

There are cf. identifications of *Ajuga* (1 sample), *Scutellaria* (10), *Origanum* (1), and identification of *Stachys*, *Prunella* and *Mentha* pollen types. All are rarely recorded in low frequency.

152. SOLANACEAE

There is a single record for this family, that of a single Devensian sample reported to contain rare *Solanum dulcamara* seeds.

154. SCROPHULARIACEAE

This family is represented by macroscopic remains and pollen, the latter more infrequently.

Macroscopic remains

Seeds of *Veronica* (3 samples), *Bartsia* (1) and *Rhinanthus* (10) have been recorded. Several species have also been identified, including *Linaria vul-*

garis (3 samples), *Pedicularis lanata* (1), *P. hirsuta* (1), *P. palustris* (9). The cf. identifications include *Veronica anagallis-aquatica* (1 sample) and *V. spicata* (1 sample). Most of these records are Devensian.

Pollen

There are rare pollen records for Scrophulariaceae (2 samples), *Scrophularia* cf. *nodosa* (1), *Bartsia* (1) and *Rhinanthus* type (2). The last is at the same site (Block Fen) as the macroscopic record.

161. LENTIBULARIACEAE

There is a single Late Wolstonian pollen record for *Utricularia*.

163. PLANTAGINACEAE

Both European genera of this family are represented by macroscopic remains and pollen, *Plantago* much more richly than *Littorella*.

Plantago

Capsules and seeds identified to the genus have been identified in 15 Devensian samples and 71 pollen samples stretching back to the Late Anglian.

P. major has been identified from seeds in seven samples and *P.* cf. *major* pollen in four samples, low in frequency and back to the Late Anglian.

P. major/media. This pollen taxon has been identified in a large number of pollen samples (138), back to the Early Pleistocene. The frequencies are usually low, up to 10 per cent, but sometimes rise to 15 per cent. Cf. *P. major/media* type is recorded in a further 19 samples.

P. coronopus pollen is recorded in 15 Devensian pollen samples in low frequencies, and a fruit of *P.* cf. *coronopus* in a further Devensian sample.

P. maritima capsules and seeds have been recorded in five Devensian samples, and *P.* cf. *maritima* in a further 10 Devensian samples. The pollen of *P. maritima* has been found in a large number of samples (122), with frequencies up to 19 per cent and back to the Early Pleistocene. *P. maritima* type pollen has been recorded in 13 further Devensian samples.

P. media pollen has been recorded in 17 samples back to the Early Wolstonian, and *P.* cf. *media* pollen in a further 16 samples within the same time range. Capsules and seeds identified to *P. media/maritima* have been recorded in two Wolstonian samples.

P. lanceolata pollen has been recorded in 21 samples, mostly Devensian.

The record of *Plantago* and the species shows that the five British species were cold stage plants, *P. maritima* and *P. major/media* being the most widely represented taxa.

Littorella

There are few records for this genus. Macroscopic remains have been identified in two samples, *Littorella* in a Devensian sample and *L. uniflora* in an Early Anglian sample. *Littorella* pollen has been found in nine samples, Devensian, Munsterian and Pre-Gortian.

164. CAPRIFOLIACEAE

There are few records for this family. The woody seeds of *Sambucus nigra* (4 samples), cf. *S. nigra* (2), *S.* cf. *racemosa* (1) and *S. nigra/racemosa* (1) have been rarely recorded in the Devensian and Wolstonian; they are often considered reworked from earlier temperate stages. There are two records of cf. *Sambucus* pollen from the Middle Devensian and two of *Lonicera xylosteum* from the Devensian Chelford Interstadial. The latter is associated with coniferous forest in Scandinavia and is likely to be contemporary.

166. VALERIANACEAE

Valerianella and *Valeriana* have both been identified in cold stage floras, the latter much more commonly.

Valerianella

Rare fruits of *V. dentata* have been recorded in two Middle Devensian samples.

Valeriana

Fruits of the following taxa have been identified, with ages back to the Early Pleistocene: *Valeriana* (3 samples), *V. dioica* (3), *V.* cf. *dioica* (2), *V. officinalis* (5), *V.* cf. *officinalis* (1) and *V. dioica/officinalis* (3). The pollen record, back to the Early Pleistocene, supports the presence of the two species of *Valeriana* in the cold stage flora. Apart from *Valeriana* records (10 samples), there are records for *V. dioica* (5 samples), cf. *V. dioica* (1) and *V. officinalis* (72).

167. DIPSACACEAE

Succisa and *Scabiosa* have both been identified in cold stage floras, the former by pollen only.

Succisa

Pollen identified as *Succisa* has been recorded in 40 samples. Pollen identified as *S. pratensis* has been recorded in a further 21 samples. Both these taxa are present in very low frequencies, back to the Early Pleistocene.

Scabiosa

There are several Devensian records for taxa of this genus, including *Scabiosa* (1 sample), *S. columbaria* (6) and cf. *S. columbaria* (2). Pollen of *Scabiosa* has been recorded in 10 samples (Devensian, Early Pleistocene), and that of *S. columbaria* in 4 samples (Devensian, Late Wolstonian).

168. CAMPANULACEAE

Campanula and *Jasione* are both represented in the flora, the former much more widely and abundantly.

Campanula

Seeds of *Campanula* are recorded in 10 Devensian samples, with cf. *Campanula* in a further three Devensian samples. Seeds of *C. glomerata* were identified an an Early Pleistocene sample and of *C.* cf. *patula/rotundifolia* in a Devensian sample. But the most abundant species is the very variable *C. rotundifolia* (60 samples; cf. 7 samples), in sediments back to the Early Pleistocene. Pollen of Campanulaceae is also widely recorded (37 samples, Devensian, Early Wolstonian), with *Campanula* (28 samples) and *Campanula* type (4 samples) in the Devensian, all in very low frequency.

Jasione

There are records for pollen of *Jasione* (6 samples) and cf. *Jasione* (5 samples), all Devensian and in low frequency.

169. COMPOSITAE

This family has a large number of taxa recorded in cold stage floras, represented by many genera and species, the latter mainly identified from macroscopic remains. Twenty-two genera are listed and over 30 species.

Macroscopic remains

The records go back to the Early Pleistocene. Many species identified have records in only one or two samples. The most widely represented genus is *Taraxacum* (28 samples), with records also for *T. officinale* agg., and several *Taraxacum* sections. *Bidens tripartita*, *Achillea millefolium*, *Matricaria perforata*, *Carduus*, *Cirsium*, incuding *C. arvense*, and *Leontodon autumnalis* are also better represented than many other genera or species. *Artemisia* and cf. *Artemisia* achenes are recorded in six Devensian samples.

Pollen

There are a large number of records of taxa identified to the family or subfamily: Compositae (159 samples), Compositae Liguliflorae (354), Compositae Tubuliflorae (266), Compositae less Compositae Liguliflorae (53). The percentages may rise above 30 per cent tlp. The records go back to the Early Pleistocene. *Artemisia* is the most widely represented genus (425 samples) (Figure 8.5), with percentages sometimes over 15 per cent tlp. *Centaurea nigra* is recorded in 22 samples, and *Saussurea alpina* in seven samples. Several pollen types are present, the more numerous of which are *Solidago* type, *Aster* type, *Anthemis* type, *Achillea* type, *Matricaria* type, *Cirsium* type and *Taraxacum* type.

The number of taxa of Compositae represented in the flora must be not only in part a result of the possibilities of distinguishing taxa, but also a reflection of the substantial nature of the seeds and pollen as well as the evident real diversity of the flora. The pollen diversity is notable, considering most genera are entomophilous, an exception being the anemophilous *Artemisia*, the most widely represented by far of the generically-identified taxa.

170. ALISMATACAEAE

Three genera of this family are represented in the flora, *Sagittaria*, *Alisma* and *Damasonium*. *Alisma* is the most commonly recorded.

Macroscopic remains

A number of taxa have been distinguished by their achenes or embryos, including Alismataceae (5 samples), *Alisma* (11), *A. plantago-aquatica* (27), *Sagittaria sagittifolia* (9), *S. natans* (1) and *Damasonium alisma* (4). The last is considered a notable southern species in the cold stage flora. The records for macroscopic remains of the family stretch back to the Early Pleistocene.

Pollen

There are records of cf. *Sagittaria* (9 samples) in the Devensian, and *Alisma* (29 samples) and *Alisma* type (7 samples) in cold stages back to the Early Pleistocene.

The family is consistently represented in the flora. The plants require habitats of shallow water and damp ground, presumably common habitats in the cold stages.

171. BUTOMACEAE

Butomus is represented by a single find of a fruit of *B. umbellatus* at a Devensian site and seven records of *Butomus* pollen in low frequency at Devensian sites. *Butomus* appears to have been a member of the aquatic flora during the Devensian.

172. HYDROCHARITACEAE

Seeds of *Hydrocharis morsus-ranae* have been recorded (2 samples) at two sites, Devensian and Late Beestonian. The former record may be of derived origin (see West *et al.* 1974). Pollen of *Hydrocharis* (3 samples) has been recorded at Beetley in the Early Devensian and Chelford Interstadial. A small number of leaf spines identified as those of *Stratiotes aloides* have been identified (6 samples) in Anglian and earlier cold stage sediments. The records indicate the possibility that the two species mentioned were members of the cold stage flora.

173. SCHEUCHZERIACEAE

There is a single record for *Scheuchzeria palustris* from the Middle/Late Devensian Lea Valley site at Ponders End.

175. JUNCAGINACEAE

There are records of fruits of *Triglochin maritima* in five samples from Devensian sites, and one record for pollen ascribed to cf. *Triglochin* at a Devensian site. The species is included in the list of obligate halophyte cold stage plants by Bell (1969).

177. POTAMOGETONACEAE

This family is very well represented in the cold stage flora by *Potamogeton* and *Groenlandia*. While the pollen identifications are at a generic or sub-generic level, a remarkable number of species identifications have been made on the basis of the fruit-stone morphology.

Macroscopic remains

The genus *Potamogeton* has been identified in 120 samples back to the Early Pleistocene, some samples having numbers of fruitstones. Nineteen species have been identified, many with additional cf. identifications. The most widely recorded species are: *P. natans* (28 samples; cf. 7 samples), *P. gramineus* (18 samples; cf. 2), *P. alpinus* (28 samples; cf. 1), *P. praelongus* (24 samples; cf. 2), *P. berchtoldii* (13 samples; cf. 9), *P. crispus* (30 samples), *P. filiformis* (80 samples; cf. 4), *P. vaginatus* (13 samples; cf. 7), *P. pectinatus* (28 samples; cf. 4). According to Polunin (1959) most of these species are recognised as reaching the Arctic. But *P. crispus* and *Groenlandia densa* (53 samples, Devensian only) have a more southern distribution, not reaching the Arctic Circle in Scandinavia. In addition to these species there are fewer records for many other species.

Pollen

There are records of pollen taxa *Potamogeton* (121 samples), *Potamogeton* type (2 samples) and *Potamogeton* subg. *Potamogeton* (1 sample). These pollen records extend back to the Anglian.

Potamogeton was clearly a consistent member of the aquatic flora of the cold stages, with a variety of species. Many well-represented species are common in the Arctic now, but there are also present species of more southern distribution.

181. ZANNICHELLIACEAE

Achenes of *Zannichellia palustris* (57 samples) have been recorded in cold stage floras extending back to the Beestonian. Some samples contain a large number of achenes. *Z. palustris* appears a consistent member of the cold stage flora, though its present distribution, while circumboreal, is north to the subarctic, not arctic.

182. NAJADACEAE

There are rare records of *Najas marina* (2 samples, Devensian), *N. flexilis* (3 samples, Devensian, Munsterian, Pre-Gortian), and *N.* cf. *flexilis* (1 sample, Early Pleistocene). The status of these two species in the flora is uncertain. *N. flexilis* is a possible member.

183. LILIACEAE

There are not many macroscopic records of this large family. *Allium schoenoprasum* has the best record, with seeds in 16 samples from many Devensian sites and in one from the Cromer complex cold stage. There is also one Devensian record of *Allium*. *A. schoenoprasum* appears a consistent member of the cold stage flora; it is at present circumboreal, north to the Arctic, often on gravel floodplains (Plate 8B). Pollen of Liliaceae taxa has been rarely recorded. The finds include Liliaceae (2 samples), cf. *Veratrum/Fritillaria* and cf. *Lloydia serotina* (1 sample each, Devensian), *Allium* (1 sample, Midlandian), and *Allium* type (1 sample, Devensian).

188. IRIDACEAE

There are only two records for this family, one Early Anglian record of of a seed of *Iris pseudacorus* and one pollen record of *Iris* in the Devensian, making the status of this family in the flora very uncertain.

189. JUNCACEAE

Both European genera of this family, *Juncus* and *Luzula*, are represented by seeds only.

Juncus

Juncus has been identified in 59 samples, of age extending back to the Early Pleistocene. Nine species have been identified, most occurring only in a few samples. *J. bufonius* is recorded in the highest number samples (11). *J. gerardii*, considered an obligate halophyte by Bell (1969) is recorded in six samples, including a Late Devensian sample with a very high number of seeds.

Luzula

Seeds of *Luzula* are recorded in 25 samples of Devensian and Midlandian age. Also recorded are *L. spicata* (2 samples; cf. 1) and *L.* cf. *multiflora* (1 sample).

The abundance of *Juncus* records is probably related to the preference of many species for damp and open habitats, where preservation is more likely, while the *Luzula* species have a preference for drier and open habitats.

193. GRAMINEAE

This is one of the best represented families in the flora, with abundant records of both macroscopic remains and pollen.

Macroscopic remains

Remains of Gramineae, mostly caryopses or fragments of flowers, have been recorded in 120 samples, in age extending back to the Early Pleistocene. Many samples contain a good number of examples. A few genera, each with few records, have been identified: *Festuca, Poa, Glyceria, Agrostis* and *Alopecurus*. The species have been more difficult to identify. There are three specific identifications: *Festuca rubra, F. halleri* and *Anthoxanthum odoratum*, with cf. identifications of several other species, including three species of *Poa* and *Elymus repens*. In terms of numbers of taxa and numbers of samples with their remains, *Festuca* and *Poa* are the most recorded genera.

Pollen

Gramineae is the best represented pollen taxon in the cold stage flora, occurring in 711 samples, usually in considerable frequencies, extending back to the Early Pleistocene (Figure 8.2). *Glyceria* type is the only other pollen taxon recorded (11 samples, Devensian).

196. LEMNACEAE

Lemna has few records. Seeds have been recorded rarely, the finds, all Devensian, including *Lemna* (1 sample), *L. trisulca* (1), *L.* cf. *trisulca* (2), *L. minor* or *trisulca* (1). Pollen of *Lemna* (3 samples) and cf. *Lemna* (2) has been recorded in very low frequency, again in the Devensian. With these rare occurrences of both seeds and pollen of *Lemna* in the Devensian, it seems probable that *Lemna* was a member of the flora in that stage. Polunin (1959) includes both species mentioned in the Arctic flora.

197. SPARGANIACEAE

Fruitstones and pollen of *Sparganium* have been commonly recorded in cold stage floras.

Macroscopic remains

The genus has been identified in 14 samples. Four species have also been recognised, of which the most common is the variable *S. erectum* (27 samples, Devensian, Wolstonian). Also recorded in lower numbers are *S. emersum* (3 samples), *S. angustifolium* (4) and *S. minimum* (5), all Devensian except for one Pre-Gortian record.

Pollen

Records of pollen of *Sparganium* (145 samples) and *Sparganium* type (89 samples) extend back to the Early Pleistocene. The frequencies are usually low. *S. erectum* type pollen has been recorded in two Devensian samples.

These records suggest that *Sparganium* species were regular members of the aquatic and marginally aquatic flora throughout cold stages.

198. TYPHACEAE

Seeds and pollen of *Typha* and its two species have a substantial record in the floras.

Macroscopic remains

Seeds of the genus have been identified in 22 samples, extending back to the Early Pleistocene. Identification of the species is more difficult. *T. angustifolia* (3 samples, cf. 2) and *T. latifolia* (7 samples, type 1) have been recorded.

Pollen

Pollen of *Typha* (3 samples) and *T. angustifolia* type (1 sample) has been recorded, but the most abundantly recorded *Typha* taxon is *T. latifolia*, the tetrads of which have been found in 88 samples, usually in low frequency, extending back to the Early Pleistocene.

As with *Sparganium*, the record for *Typha* suggests that *T. latifolia* and probably *T. angustifolia* (since *Sparganium type* pollen may include *Typha*) were consistent members of the cold stage flora.

199. CYPERACEAE

Eight genera and a large number of species have been identified on the basis of abundant macroscopic remains: *Scirpus*, *Blysmus*, *Eriophorum*, *Eleocharis*, *Cyperus*, *Cladium*, *Rhynchospora* and *Carex*. The family is stenopalynous, and the record for Cyperaceae pollen is very substantial. It is second only to the Gramineae record (Table 8.5), with 666 samples containing the taxon in often high frequencies, extending back to the Early Pleistocene (Figure 8.3). There is in addition one pollen record for *Cladium* (Pre-Gortian) and one for *Rhynchospora alba* (Devensian).

The more detailed record of macroscopic remains (utricles, nuts) also extends back to the Early Pleistocene. It can best be reviewed genus-by-genus, first noting that the Cyperaceae taxon is represented in 27 samples.

Scirpus

The genus is recorded in 25 samples. The most common species recorded is *S. lacustris* (64 samples), with subspp. *tabernaemontani* (3 samples) and *lacustris* (7 samples) also identified. *S.* × *carinatus* and *S. setaceus* are each recorded in one sample, and there are cf. identifications of two other species, each only in one or two samples.

Blysmus

B. compressus and *B. rufus* have each been found in two samples, the former Devensian and the latter Devensian and Early Wolstonian.

Eriophorum

The genus has been recorded in one Devensian sample, and *E. angustifolium* in three Devensian samples. *E. vaginatum* records are based largely on the occurrence of the sclerenchymatous spindles, which have been found in the Early Devensian (25 samples, 2 sites) and in one Early Pleistocene sample.

Eleocharis

The genus has been identified in 13 samples, extending in age back to the Early Pleistocene. The most frequently recorded species is *E. palustris* (65 samples; cf. 13), with subspp. *vulgaris* (3 samples) and *palustris* (5 samples) also identified. *E. uniglumis* is recorded in 22 samples. *E. quinqueflora* has been identified in five samples, and there are single records for *E. parvula, E. multicaulis* and *E. carniolica*.

Cyperus

Cf. *Cyperus* has been recorded in one Early Pleistocene sample, and *C. longus* in two samples (Anglian, Devensian).

Cladium

There are five records for the woody nuts of *C. mariscus*. Certain of these (Wretton, Tottenhill) are considered to be derived from underlying and neighbouring temperate stage sediments. The status of *Cladium* in the cold stage flora is uncertain.

Rhynchospora

There are two ? records of *Rhynchospora* from the Devensian Lea Valley sites (pollen at one Devensian site, see above).

Carex

Records of the genus are very abundant, extending in time back to the Early Pleistocene. The taxa include *Carex* (230 samples), biconvex *Carex* nuts (33 samples), and trigonous *Carex* nuts (26 samples). It is remarkable that over 25 species have been identified, in addition to a number of cf. identifications. The species recorded most commonly is *C. rostrata* (32 samples), and the *C. aquatilis/bigelowii* complex is also well represented (22 samples). *C. flacca, C. flava* and *C. nigra* have 6 to 8 samples each. But the majority of taxa are only identified in a few samples.

A crucial point is the interpretation of the abundance of records for Cyperaceae and *Carex*. The abundance of macroscopic remains must reflect the preference of many species of the genus for damp or aquatic habitats, which favour preservation of their remains. The high frequencies of Cyperaceae pollen may also result from the same preference. But the possibility of a regional vegetation with Cyperaceae also has to be considered, a problem discussed in Chapter 12.

Other taxa recorded

Nyssa

Pollen of *Nyssa* has been recorded in Late Wolstonian sediments at Selsey; it is considered derived from neighbouring Tertiary sediments.

'Type X'

This tricolpate reticulate pollen grain of uncertain affinity (see Turner 1970), has been recorded in cold stage sediments (Wolstonian) post-dating Hoxnian temperate stage sediments, where it occurs with forest pollen assemblages. It is considered derived.

Sphagnum

Spores of *Sphagnum* occur in a large number of samples (319), usually in very low frequencies. Higher frequencies occur in the Devensian Chelford Interstadial and in the Early Pleistocene.

Characeae

Oospores of several taxa in this family have been recorded throughout cold
stages, extending in age back to the Early Pleistocene. These taxa include
Characeae (55 samples), *Chara* (91), *Nitella* (44), *Tolypella* (2). Large
numbers of oospores occur in some samples.

Reworked Quaternary spores and pre-Quaternary palynomorphs

These are recorded, in some samples in high frequencies, in the cold stages
back to the Early Pleistocene (total of 202 samples). The frequencies are
valuable in demonstrating the reworking which is associated with the deposition of cold stage assemblages, and of the consequent problem of determining which taxa are contemporary and which reworked. *Megaspores* (23
samples) are in the same reworked category.

8

The representation of taxa in the fossil record

In this chapter consideration is given to two aspects of the representation of taxa in the fossil record. First, problems relating to identification, briefly mentioned in Chapter 6, are discussed in more detail. Secondly, the facts of the representation of taxa (macros and pollen) in the data base, are summarised, in preparation for interpretations given in later chapters. Except where stated, the numbers of taxa given include all taxa identified and listed in the data tables.

Taxonomic problems and uncertainties

In any analysis, the number of identified taxa is related to the potential identifiability of family, genus, species or subspecies, the preservation capability of the remains in question, and the taphonomy of the fossil assemblage (see Chapter 3). All these affect the significance of the assemblage for the reconstruction of the flora, with the fossil flora being a selected portion, sometimes small, sometimes large, but always biased, of the flora of the time.

A major difficulty of identification to species level is the variability of many taxa of common cold stage genera, coupled with the difficulty of making comparisons with reference material from the continental European flora. Species and subdivisions of species in the data tables mostly follow *Flora Europaea*, but 'critical' species (as e.g. in *Salix*, *Draba*), common in the cold stage flora, handicap the enumeration of the fossil flora.

A further problem is the identification of hybrid taxa. Such taxa are likely to have been present in the disturbed habitats and variable climates and micro-climates of cold stage environments, leading to introgression and evolution of ecotypes, as with the hybrid zones discussed by Rieseberg & Wendel (1993).

These difficulties involve taxonomy and morphology. Rarely, morphology indicates a particular variant, as with the recognition of an extinct taxon related to *Picea glauca* from cone morphology of fossils from full-glacial sediments in southern U.S.A. (Jackson & Givens 1994), an interpretation which has helped to clarify the climatic meaning of the associated assemblage.

But there is also the unrecognisable variability which may result from the presence of biotypes within a taxon. Physiological variability and its description are an unknown for most taxa, and cannot yet be recognised in a fossil assemblage. An exception is *Saxifraga oppositifolia*. The genetic diversity and ecological variants of living populations of this species have been described by Crawford (1997), Crawford & Abbott (1994) and Crawford, Chapman & Smith (1995). If such could be recognised in the fossil record of *S. oppositifolia*, it would be invaluable for the study of past environments and climates, as well as for the evolution and taxonomy of the species.

These problems, involving lack of knowledge of biotypes, variation and hybridity, certainly make for difficulty in interpreting past conditions of climate and habitat from the fossil record. They have to be borne in mind in making such interpretations.

Apart from these difficulties related to identification, preservation capability will bias the content of the fossil assemblage, via taphonomy and the hardiness of the plant part. Variation in the ability of pollen taxa to withstand taphonomic and post-depositional processes is well known (e.g. Havinga 1984), and macroscopic remains are also subject to similar forces, including floatability, dispersal mechanisms, and seed or fruit size.

The representation of taxa

Families, genera and species

A broad conspectus of the cold stage flora of higher plants as a whole can be based on figures taken from the TAXASORT table for numbers of families, genera and species identified. For a comparison with the present European flora, the figures in parentheses indicate the relevant numbers of native taxa in the *Flora Europaea* given by Webb (1978).

Families: 78 families, plus 18 represented by remains likely to be reworked and 6 more doubtfully so (164).

Genera: 235 genera, plus 25 likely to be reworked and 6 doubtful (1340).

Species: 364 species identified (labelled 'a' and 'da' in the use field of TAXA-SORT), 297 on the basis of macroscopic remains only, 37 on the basis of pollen or spores and 36 on the basis of both macroscopic remains and pollen or spores (10,200); Webb (1978) also gives figures for territories covered by *Flora Europaea*, including Britain (1700–1850), France (4300–4450), Germany (2600–2750) and Sweden (1600–1800).

Although a comparison of the cold stage flora with the *Flora Europaea* numbers is of minor significance since cold stage conditions must have been very different from present conditions, it may be of interest since the cold stage flora, though a partial representation of the flora of the time, is antecedent to the present flora. Of the 24 most species-rich families in *Flora Europaea* listed by Webb (1978) (Table 8.1), 13 are represented in the cold stage flora by higher numbers (>4) of species (e.g. Caryophyllaceae FE (*Flora Europaea*) 655, CS (cold stage) 30; Compositae FE 1326, CS 30; Cyperaceae FE 260, CS 42). On the other hand, some families rich in species have had no or few species identified in the cold stage flora (e.g. Rubiaceae, FE 234, CS 0; Gramineae FE 880, CS 3; Orchidaceae FE 114, CS 0), by reason of absence, the difficulty of identification of species via pollen, spores or macroscopic remains (e.g. Rubiaceae, Gramineae), the low preservation potential of their seeds or pollen (Orchidaceae), or finally the lack of local sedimentary processes which encourage preservation of fossils.

Table 8.1 compares the *Flora Europaea* numbers with the cold stage numbers in the families. As might well be expected, the list is biased towards species which are aquatic or favour damp ground (genera of Polygonaceae, Potamogetonaceae, Juncaceae, Cyperaceae). *Potamogeton* especially is well represented, probably a result of its aquatic habitat and potential identification characters, but perhaps also reflecting a wide ecological tolerance in the species and possibly reduced competition in aquatic conditions. At the other extreme is the small number of Gramineae species identified, though it is the most commonly occurring pollen taxon in the samples, a consequence of difficulty of identification of caryopses even though Gramineae macroscopic remains are commonly recorded. Intermediate are herbaceous families which characterise open plant communities, such as Caryophyllaceae, Cruciferae and Compositae, with a fair number of species recorded in the cold stage flora. In comparison the representation of tree and shrub taxa is low.

In respect of the number of species identified, Table 8.2 relates the number of species ('a' and 'da' in TAXASORT) identified to the number of

Table 8.1. *Comparison of species numbers*

Family	*Flora Europaea* 24 most species-rich families (Webb 1978)	Cold stage flora species ('a' or 'da' in TAXASORT table)
Compositae	1326	30
Gramineae	880	3
Leguminosae	840	6
Caryophyllaceae	655	30
Cruciferae	649	18
Scrophulariaceae	515	4
Labiatae	452	8
Umbelliferae	431	11
Liliaceae	371	1
Ranunculaceae	310	20
Rosaceae	263	17
Boraginaceae	262	1
Cyperaceae	260	42
Rubiaceae	234	0
Campanulaceae	211	3
Chenopodiaceae	158	8
Plumbaginaceae	146	1
Saxifragaceae	129	6
Dipsacaceae	128	2
Euphorbiaceae	118	2
Orchidaceae	114	0
Crassulaceae	107	1
Polygonaceae	104	20
Primulaceae	101	5
Other families		
Salicaceae	81	10
Ericaceae	45	6
Plantaginaceae	36	6
Potamogetonaceae	23	20
Juncaceae	54	10

samples, macro and pollen, in which the species have been identified. Nearly half the species identified macroscopically have only been found in one or two samples, usually in very low frequency, a proportion which has to be borne in mind in any analysis of the figures. The same can be said of the pollen identifications of species, about a third being in the same categories. The question then arises: should any reconstructions be based on the whole list of species (with all the problems of identification), or on the more commonly occurring species? Study of one or the other list may lead to the same conclusions, with reasonable reconciliation, relating, for example, to

Table 8.2. *Occurrence of species ('a'or 'da') in samples*

Sample numbers occurrence class	No. species macros	No. species pollen/spores
1	110	23
2	47	6
3	37	5
4	19	5
5	19	3
6	15	3
7	6	4
8	5	2
9	9	0
10–14	18	1
15–20	10	4
21–50	27	14
51–100	12	2
>100	0	1
Total	334	73

variable habitats. Or there may be problems of reconciliation, possibly a result of, for example, biotypic variation of species. Such difficulties are discussed in a later chapter.

These figures show the kind of bias inherent in a fossil assemblage and the problem of producing a balanced view of the a flora which could be used to indicate past climates and environments.

Variation of number of taxa in time

An analysis of the number of taxa recorded in time (TAXATIME) will give a measure of the scale of knowledge of the flora at particular times. Table 8.3 gives the number of macro and pollen or spore taxa identified in the stratigraphical subdivisions used in the data base. The number identified is obviously related to the number of sites studied (Table 5.3), with the number greatest in the most recent cold stage, the Devensian, fewer in the Wolstonian, and far fewer in earlier times. The number in the Devensian relates to the large number of river floodplain organic sites recorded (especially Middle and Late Devensian), many of which have large floras favoured by their taphonomy. In earlier times such sites are much rarer. In addition to this rarity, in the Lower Pleistocene the poor floras are related

Table 8.3. *Number of taxa in stages and substages*

Stage, substage	Macroscopic remains	Pollen or spores
l De	353	134
m l De	152	21
m De	446	233
m De UW	187	91
e m De	37	55
e De Br	–	41
e De Ch	54	99
e De	167	131
De	248	89
l Wo	77	86
e Wo	124	69
Wo	142	63
Wo ?	24	–
l An	79	57
m An	38	–
e An	84	37
l Be	36	b 30
		a 28
Cccs	92	–
Be	55	46
PrePa d	21	34
b–d	–	17
c	14	86
b	1	30
a	23	54
PrePa	–	22
Ba	–	50
Lu 4c	–	28
Lu 4b	–	23
Lu 2/4	9	18
Th	–	31
Lu 1/2	11	35
m Mi	26	23
e m Mi	19	18
e Mi Ag	21	7
Mu	20	23
PreG	35	29

to taphonomy of the assemblages found in the marine sediments, with pollen taxa numbers far exceeding macro taxa, unlike the Devensian.

Representation of identified taxa

Here we consider the representation of particular macro and pollen taxa, in terms of the number of samples in which the taxa occur, and in terms of the frequency of the taxa in the samples.

Taxa and sample numbers

Using the TAXASORT table as the data source, an order of frequency of occurrence in the samples of macro and pollen taxa can be drawn up. Table 8.4 shows the frequency of occurrence of the commonest 50 macro taxa in this ordination, and Table 8.5 the commonest 50 pollen taxa. Each of these tables also gives the representation of the taxon by pollen in the macro table and by macros in the pollen table, so showing which taxa are represented by macros, which by pollen and which by both. Since the macro identifications are often at the species level, while the pollen identifications are often at a genus or higher level, a relation between the occurrence of macros and pollen of a connected species and genus will not be shown: for example, *Helianthemum canum* occurs in 45 macro samples but no pollen samples, while *Helianthemum* pollen has been recorded in seven macro samples and 97 pollen samples.

The most commonly occurring macro taxa (in >100 samples) are *Carex* spp., *Ranunculus Batrachium*, Gramineae and *Salix*, while the most commonly occurring pollen taxa are Gramineae, Cyperaceae (matching the *Carex* frequency), *Pinus* and *Betula*. A comparison of the occurrences shows that Gramineae and Cyperaceae (*Carex*) are very well represented by both macros and pollen, as is *Salix*. On the other hand, for example, *Pinus* and *Artemisia* are well represented by pollen, but have very few macro records, reflecting the wide distribution of these taxa as pollen, though hardly recorded as macro remains. Table 8.6, extracted from the TAXASORT table, lists the taxa which are represented by macros (>5 samples) and pollen (>2) samples. A study of the complete extracts from the TAXASORT table, on the lines of Tables 8.4 and 8.5, shows clearly how and to what extent taxa are represented in the samples.

The variability of representation evident in these tables underlines the difficulty of interpreting the flora in terms of vegetation or plant communities. Remains of each taxon will have their own particular character in

Table 8.4. *The 50 commonest macro taxa present in samples*

Use	Name	Taxon	M/P	No. macro	No. pollen
	Carex spp.	63	m	191	0
m	*Ranunculus Batrachium*	223	m,p	189	7
m	Gramineae	92	m,p	120	711
	Salix	261	m,p	105	391
a	*Hippuris vulgaris*	101	m	97	0
a	*Potentilla anserina*	174	m	87	0
m	*Chara*	214	m	86	0
a	*Potamogeton filiformis*	142	m	80	0
a	*Armeria maritima*	127	m,p	72	41
	Potamogeton	259	m,p	70	121
a	*Selaginella selaginoides*	197	m,p	69	33
a	*Scirpus lacustris*	224	m	64	0
a	*Eleocharis palustris*	64	m	62	0
a	*Campanula rotundifolia*	9	m	60	0
a	*Ranunculus sceleratus*	164	m	58	0
a	*Zannichellia palustris*	212	m	57	0
	Viola	264	m	55	0
	Characeae	216	m	55	0
a	*Betula nana*	7	m,p	54	15
	Cruciferae	51	m,p	53	251
a	*Groenlandia densa*	141	m	53	0
a	*Urtica dioica*	317	m	48	0
a	*Menyanthes trifoliata*	120	m,p	47	6
	Potentilla	176	m,p	46	90
	Potamogeton spp.	149	m	46	0
a	*Helianthemum canum*	34	m	45	0
	Rumex	134	m,p	44	88
m	*Nitella*	215	m	44	0
a	*Salix herbacea*	187	m,p	42	4
	Ranunculus subg. *Ranunculus*	239	m	41	0
a	*Diplotaxis tenuifolia*	57	m	41	0
	Juncus	103	m	39	0
a	*Thalictrum flavum*	166	m	38	0
	Draba	266	m	37	0
a	*Draba incana*	49	m	37	0
	Betula	3	m,p	35	582
da	*Polygonum aviculare*	132	m,p	35	9
	Salix spp.	186	m	34	0
a	*Ranunculus flammula*	161	m	34	0
	Umbelliferae	206	m,p	34	289
	Carex	255	m	34	0
a	*Myriophyllum spicatum*	100	m,p	33	50
	Carex-biconvex	876	m	33	0
da	*Rumex acetosella* agg.	136	m	32	0
da	*Linum perenne* agg.	119	m	32	0
a	*Carex rostrata*	254	m	32	0
a	*Polygonum viviparum*	130	m,p	31	1
	Caryophyllaceae	21	m,p	30	405
a	*Potamogeton crispus*	140	m	30	0
	Betula pendula or *pubescens*	225	m	29	0

Table 8.5. *The 50 commonest pollen/spore taxa present in samples*

Use	Name	Taxon	M/P	No. pollen	No. macro
m	Gramineae	92	m,p	711	120
m	Cyperaceae	358	m,p	666	27
	Pinus	422	m,p	643	2
	Betula	3	m,p	582	35
m	*Artemisia*	617	m,p	425	2
	Caryophyllaceae	21	m,p	405	30
	Salix	261	m,p	391	105
	Filicales (Polypodiaceae)	566	m,p	367	2
	Compositae Liguliflorae	926	p	354	0
m	*Sphagnum*	966	p	319	0
	Thalictrum	169	m,p	315	7
	Rubiaceae	930	p	304	0
	Umbelliferae	206	m,p	289	34
	Alnus	672	m,p	288	3
	Picea	423	m,p	283	7
	Campanula	10	m,p	282	8
	Compositae Tubuliflorae	927	p	266	0
	Cruciferae	51	m,p	251	53
	Chenopodiaceae	818	m,p	227	1
	Pre-Quaternary palynomorphs	950	p	188	0
	Corylus	932	p	177	0
	Calluna	934	p	176	0
m	*Empetrum*	607	m,p	169	16
	Ranunculaceae	941	p	166	0
	Ericales	974	p	164	0
	Compositae	620	m,p	159	8
	Ranunculus	159	m,p	156	14
	Sparganium	412	m,p	145	10
	Armeria	128	m,p	139	24
	Plantago media/major	929	p	138	0
m	*Botrychium*	945	p	135	0
	Quercus	912	m,p	129	4
	Juniperus	675	m,p	122	1
a	*Plantago maritima*	124	m,p	122	5
	Potamogeton	259	m,p	121	70
	Rosaceae	942	p	114	0
	Carpinus	931	p	102	0
	Polypodium	982	p	97	0
	Helianthemum	35	m,p	97	7
	Potentilla	176	m,p	90	46
	Sparganium type	947	p	89	0
	Rumex	134	m,p	88	44
a	*Typha latifolia*	205	m,p	88	7
	Ulmus	968	p	81	0
	Caltha	989	p	81	0
a	*Valeriana officinalis*	338	m,p	72	5
	Plantago	549	m,p	71	14
	Solidago type	1143	p	64	0
	Hippophae	1073	p	63	0

Table 8.6. *Taxa well-represented by macros and pollen*

	Number of samples	
	macro (>5) (ex 386)	pollen (>2) (ex 720)
Gramineae	120	711
Cyperaceae, *Carex*	218	666
Betula	35	382
Caryophyllaceae	30	405
Salix	105	391
Thalictrum	7	315
Umbelliferae	34	289
Picea	7	283
Campanula	8	282
Cruciferae	53	251
Empetrum	16	169
Compositae	8	159
Ranunculus	14	156
Sparganium	10	145
Armeria	24	139
Potamogeton	70	121
Helianthemum	7	97
Potentilla	46	90
Rumex	44	88
Typha latifolia	7	88
Plantago	14	71
Myriophyllum spicatum	33	50
Ericaceae	6	47
Armeria maritima	72	41
Menyanthes	6	41
Polygonum aviculare	7	34
Selaginella selaginoides	69	33
Alisma	11	29
Rumex acetosella	25	26
Myriophyllum alterniforum	16	25
Betula nana	54	15
Polygonum aviculare agg.	35	9
Ranunculus Batrachium	189	7
Menyanthes trifoliata	47	6
Filipendula ulmaria	17	5
Linum catharticum	8	5
Salix herbacea	42	4
Ranunculus spp.	7	4
Saxifraga oppositifolia	20	3
Linum perenne ssp. *anglicum*	23	3
Typha	18	3

Table 8.7. *Numbers of taxa and species ('a' or 'da') in the flora*

Macroscopic remains only	959 taxa	297 species
Pollen or spores only	270 taxa	36 species
Macroscopic remains and pollen or spores	113 taxa	37 species

terms of production by the parent plant, dispersal of the remains and ease of preservation. Thus the taxa listed in Table 8.6 cover a variety of habitats, terrestrial and aquatic, life forms and pollination methods. Some generalisations may be possible: for example, the pollen taxa most frequently occurring (Table 8.5) are of anemophilous plants, but entomophilous taxa are also frequent, indicating their significance in the flora. Additionally, Table 8.6 is also obviously a reflection of the ease of identification of particular taxa.

Macro and pollen representation in the samples

In any compilation of a fossil flora involving macroscopic remains and pollen or spores, with identifications at taxonomic levels from family to subspecies, there is a problem in evaluating the presence or abundance of the members of the flora either locally or regionally. Three categories of representation, all potentially contributing to our knowledge of the flora and vegetation, are given in the TAXASORT table: taxa are represented by macroscopic remains, pollen, or both. These categories are now discussed in more detail, with their possible significance for the interpretation of the flora.

Macroscopic remains

959 taxa are recorded only by their macroscopic remains, of which 297 are at the species level (Table 8.7). Macroscopic remains provide the great majority of species identifications in the flora, and are thus of great importance in interpretations of the flora. The number of samples in which species have been recorded is shown in Table 8.2. About a third have only been found in single samples. In the most abundant occurrence class, 51–100, there are 12 species (see Table 8.8), of which nine are not represented at the specific level by pollen, but are, except for *Zannichellia*, represented at generic level by pollen. These commonest species include both aquatics and terrestrial plants, the former predominating. In the lower

Table 8.8. *Species most represented in samples by*
macroscopic remains

		Sample number	
Species	Macro/pollen	Macro	Pollen
Armeria maritima	m,p	72	41
Betula nana	m,p	54	15
Campanula rotundifolia	m	60	0
Eleocharis palustris	m	62	0
Groenlandia densa	m	53	0
Hippuris vulgaris	m	97	0
Potamogeton filiformis	m	80	0
Potentilla anserina	m	87	0
Ranunculus sceleratus	m	58	0
Scirpus lacustris	m	64	0
Selaginella selaginoides	m,p	69	33
Zannichellia palustris	m	57	0

occurrence classes, terrestrial species become better represented, the change reflecting the reduced preservation possibilities of remains of terrestrial species. The terrestrial species in the higher classes (e.g. *Helianthemum canum, Diplotaxis tenuifolia, Urtica dioica, Campanula rotundifolia, Potentilla anserina*) must have been widespread, but nevertheless selectively preserved, and identified because of the hardy nature of their remains.

When there are one or few records of particular species (or taxa), a question arises about the significance of the remains and whether they may be reworked. Some help is given here by the species with few macro records, but much more abundant pollen records. For example, *Sanguisorba officinalis* has been found in one macro sample, but 22 pollen samples, *Valeriana officinalis* in five macro samples and 72 pollen samples, and *Plantago maritima* in five macro samples and 122 pollen samples. A poor macro record clearly does not mean a reworked origin. An examination of the sedimentary context and nearness of a source of reworked fossils is also necessary for any conclusions about reworking.

Pollen and spores

270 taxa are recorded only by their pollen or spores, of which 36 are at a species level. Thus most of the identifications are at the genus or family level. The greatest number of species is recorded in only one sample, as with the macroscopic remains. Species of *Plantago, Centaurea nigra, Succisa*

pratensis, Lycopodium annotinum and *Huperzia selago* are the most commonly occurring species in the samples (17–24). The genus or family occurrence numbers are far greater (see Table 8.5). They provide a far more generalised view of the flora and vegetation, with very significant indications of vegetation formations and their content.

Macroscopic remains and pollen or spores

113 taxa are recorded by both macros and pollen or spores, of which 37 are at a species level. Eight of the species are aquatic, where it might be expected that both types of remains would be well preserved, but the larger number are in fact terrestrial and include *Armeria maritima, Betula nana, Linum perenne, Rumex acetosella, Salix herbacea* and *Selaginella selaginoides*. The remaining taxa represented by both macroscopic remains and pollen or spores are are mainly at genus level, with fewer at family level.

Apart from those taxa recorded by both macroscopic remains and pollen or spores, the identification of genera or families by pollen or spores is often well supported by species identifications from macroscopic remains, as with *Potamogeton* (121 pollen samples) and the many *Potamogeton* species identified. Likewise, for example, Caryophyllaceae, Cruciferae, Leguminosae, Umbelliferae and Compositae are supported by species identifications. These relationships are clearly seen in the TAXASORT table. The balance between macro and pollen or spore representation may give guidance about reworked taxa. Thus *Corylus, Carpinus* and *Quercus* are represented by many very low frequency pollen records, but their presence is not supported by equally frequent macro records.

Frequencies of taxa in the samples

Macroscopic remains

The frequency of macro taxa in each sample is given in the MACRO table. The frequencies of taxa in samples in this table can be grouped into classes. Table 8.9 shows the number of taxa in particular frequency classes. Taxa most frequently found in samples are listed in Table 8.10.

Where numbers of remains are reported, sample totals are also given for each sample. Percentage frequencies are not usually reported in publications, but can be calculated from the MACRO table. As will be discussed, the significance of such percentages is problematical. There is little comparison with the use of percentages in pollen analysis, where numbers of fossils are high and taphonomy is usually simpler.

Table 8.9. *Number of taxa of macroscopic remains in frequency classes*

Frequency of a taxon (no. of remains) in a sample	Number of taxa
1	1969
2	779
3	373
4	258
5	171
6	141
7	109
8	97
9	92
10	63
11	58
12	55
13	38
14	37
15	35
16	29
17	26
18	26
19	15
20–24	89
25–29	60
30–34	48
35–39	31
40–44	32
45–49	18
50–59	43
60–79	38
80–99	27
100–149	42
150–199	18
200–299	16
300–399	11
400–599	12
600–799	3
800–999	3
1000+	8

Table 8.10. *Taxa of macroscopic remains*
represented most abundantly in the samples

No. of remains in a sample	Taxon (No. of samples, if >1)
>999	*Carex* spp. (2)
	Characeae (2)
	Diplotaxis tenuifolia
	Nitella
	Picea abies cf. ssp. *obovata*
	Ranunculus Batrachium
800–999	*Juncus gerardii*
	Nitella
	Ranunculus Batrachium
600–799	*Carex* spp. (2)
	Ranunculus Batrachium
400–599	*Carex* spp.
	Characeae (3)
	Diplotaxis tenuifolia
	Eleocharis palustris + *uniglumis*
	Nitella (3)
	Pinus subg. *Pinus*
	Ranunculus Batrachium
	Selaginella selaginoides
300–400	*Carex* spp. (2)
	Characeae (2)
	Ranunculus Batrachium
	Nitella
	Betula nana
	Lycopus europaeus
	Eleocharis palustris
	Polygonum aviculare
	Urtica dioica
200–299	*Carex* spp. (2)
	Characeae (2)
	Selaginella selaginoides (2)
	Ranunculus Batrachium (2)
	Ranunculus hyperboreus
	Nitella
	Hippuris
	Betula pendula or *pubescens*
	Lycopus europaeus (3)
	Cruciferae

Table 8.10 (*cont.*)

No. of remains in a sample	Taxon (No. of samples, if >1)
150–199	*Carex* spp. (2)
	Potamogeton filiformis
	Potentilla anserina
	Selaginella selaginoides (2)
	Characeae (2)
	Ranunculus Batrachium (2)
	Myriophyllum alterniflorum
	Diplotaxis tenuifolia (2)
	Nitella
	Scirpus lacustris
	Scirpus lacustris ssp. *lacustris* (2)
	Hydrocotyle vulgaris
100–149	*Potentilla palustris* (2)
	Salix herbacea
	Selaginella selaginoides (5)
	Draba cf. *incana*
	Salix
	Carex spp. (4)
	Carex-biconvex
	Armeria maritima (2)
	Polygonum aviculare agg.
	Groenlandia densa
	Potamogeton filiformis
	Zannichellia palustris
	Ranunculus Batrachium (3)
	Carex bigelowii/aquatilis
	Potamogeton praelongus
	Tolypella nidifica
	Polygonum oxyspermum agg.
	Potamogeton vaginatus
	Glyceria cf. *fluitans*
	Diplotaxis tenuifolia (2)
	Scirpus lacustris
	Scirpus lacustris ssp. *lacustris* (2)
	Hippuris vulgaris
	Betula cf.*pubescens*
	Draba spp.
	Draba incana
	Eleocharis palustris ssp. *palustris*
	Chara
	Juncus spp.

There are different kinds of representation of taxa in the samples, which may be distinguished as follows:

1. Taxa which occur in a considerable number of samples (Table 8.4).
   a. with low frequency of remains, e.g. *Menyanthes trifoliata*, *Campanula rotundifolia*.
   b. with often a high frequency of remains, e.g. *Salix*, *Ranunculus Batrachium*, *Potentilla anserina*, *Carex*.
2. Taxa which occur in a small number of samples.
   a. with a low frequency of remains, e.g. *Parnassia palustris*.
   b. with sometimes a high frequency of remains, e.g. *Picea abies* cf. ssp. *obovata*.

The factors which govern this variation are complex. They include intrinsic matters such as seed/fruit production, seed/fruit size and shape, capacity for floatation, leaf toughness and production of deciduous shrubs, and, since plant remains are sedimentary particles, matters of taphonomy such as water flow rates during dispersal and sedimentary processes affecting deposition. Variation of seed size is great (see Salisbury 1942) and so are powers of floatation (see Praeger 1913). The longest lists of species identified from macros come from sediments in river floodplains, where sedimentary conditions favour survival of remains of a wide variety of plants in transient pools of aggrading rivers. These are just the conditions, in a fluvial regime, that make understanding the taphonomy so important. In contrast, quietwater sediments of, for example, meander cut-offs may contain a far more limited set of remains, in number and in taxa.

So the question remains of the significance of the analysis described above. If a taxon occurs widely in a number of samples (1, above), it can be taken to have a wide distribution in the area concerned, even if it shows a low frequency of remains in particular samples (1a, above). If there is a very high frequency of remains (Table 8.10; 1b, above), two explanations may apply. Abundant production of, for example, fruits of *Carex* from local plant communities bordering pools may lead to high frequencies of the remains in an assemblage from a pool sediment. Or the taphonomy has positively aided the accumulation of remains of a particular taxon, as in a leaf flora with *Salix* leaves (Figure 8.1) or power of floatation has led to concentration, as seems likely with *Ranunculus Batrachium* achenes, often very abundant. The nature of the containing sediment, with evidence of stagnant or fluvial conditions, should decide which explanation may apply.

Where taxa have been recorded in a smaller number of samples (2, above), as is the case with most taxa (Table 8.2), the interpretation is less

Figure 8.1. Plant remains in a small stream in polar desert, Bathurst Island, Nunavut, Canada. Besides the visible leaves of *Salix arctica*, the extracted remains showed the presence of other taxa, including *Draba*, *Dryas integrifolia*, Gramineae, *Papaver radicatum* and *Saxifraga oppositifolia*. This is the present-day equivalent, in process terms, of a Quaternary cold stage leaf flora. 5 cm intervals on ruler.

clear. The taxa may be less widespread in the flora, growing where incorporation of remains into sediment is less likely, or the production of seeds/fruits or leaves may be lower. In a few cases, taxa occurring in few samples may be very abundant in a sample (2b, above), as with *Picea abies* cf. ssp.*obovata*. Here the sedimentary environment is interpreted as a shallow pool (Phillips 1976), and concentration may result from local abundance of the tree with a flotation effect associated with the pool.

In spite of these difficulties, the numbers of samples in which taxa are found and the frequency in particular samples do have significance for interpretation of the flora and vegetation, as suggested above, as well as providing information about the bulk of species identifications in the cold stage flora.

Pollen and spores

Frequencies of pollen and spore taxa are given in the POLLEN table. The variety of expression of frequency, in terms of the percentage base, complicates matters. The base for calculation for each sample is given in the

PLSAMPLE table. Commonly it is the land plant pollen total which is used. In the POLLEN table, frequencies are given as percentages or percentage categories. These data can be extracted from the POLLEN table for each taxon.

As with the macroscopic remains, different kinds of representation of taxa can be distinguished:

1. Taxa present in many samples.
   a. taxa with usually higher frequencies, e.g. Gramineae, Cyperaceae.
   b. taxa with variable frequencies, often low but with occasionally higher, e.g. *Pinus*, *Abies*, Cruciferae, *Helianthemum*, *Plantago major/media*, *P. maritima*, Compositae Liguliflorae.
   c. taxa with low representation in the samples, e.g. *Selaginella*, *Carpinus*, Chenopodiaceae, Rosaceae, Leguminosae, *Polemonium*, *Succisa*, *Alisma*.
2. Taxa present in fewer samples, e.g. *Taxus*, *Corylus*, *Urtica*, *Montia fontana*, *Rubus chamaemorus*.

The two anemophilous taxa in the first category, 1a, present less difficulty in interpretation. Their high frequencies are paralleled by high macro remains frequency, and their presence characterises the pollen spectra as cold stage pollen spectra.

The other categories are more difficult to interpret, because of the possibilities of reworking of pollen and spores. Judgment about reworking of taxa is complex, relying on taphonomy and ecological characters of the taxa concerned. Category 1b contains taxa whose presence is supported by the presence of macro remains (e.g. Cruciferae, *Helianthemum*); the variation in frequency must here be related to proximity of parent plant and to taphonomy. But it also contains taxa which are more abundant in underlying temperate stage sediments and where reworking is very likely. For example, *Abies* and *Alnus* pollen is usually in very low frequency and not supported by macro remains, but there are higher frequencies in Early Wolstonian sediments which overlie temperate stage sediments containing these taxa in abundance: reworking is the likely cause of the higher pollen frequencies in such cases. It is notable that several taxa in categories 1b and 1c, present in many samples, are entomophilous; though pollen production may be low, they are widely represented.

Category 1c may also include reworked taxa, such as *Corylus*, *Carpinus* and *Quercus*, temperate trees with little support for presence from macro remains. Other taxa, e.g. Leguminosae and Rosaceae, are well supported by macro remains of a variety of taxa. Similar difficulties apply to the taxa in category 2. *Urtica*, with a strong macroscopic record, is taken to be present, but the rare pollen of the temperate tree *Taxus* is considered reworked.

The behaviour of pollen and spore taxa can also be illustrated by relating percentage frequency (see above: commonly land plant pollen total; see also Chapter 5) to sample number. If the percentages are translated to the pollen percentage categories, distribution of the total of percentage categories can be related to sample number (Table 5.3) in each stage/substage, as in the graphs, Figures 8.2 to 8.10. Scan occurrences are excluded from this analysis. Such graphs summarise the importance of the taxa concerned in the cold stage pollen rain, giving an overall view of pollen representation and its variation for each taxon. Graphs for each taxon cover Devensian stadial divisions, the Devensian interstadials (UW, Br, Ch), and older stadials (e Wo, Wo, e An, Be, PrePa b–d, PrePa a and Ba). Late-glacial samples (l Wo, l An, l Be) are not used.

The distribution of numbers of samples in particular pollen categories will be affected by the facies of the sediment containing the pollen assemblages and by their taphonomy. Thus early-glacial samples appear to be more affected by redeposition from temperate stage pollen assemblages than those of the later parts of cold stages, and in the Early Pleistocene, low Cyperaceae percentages are typical of the marine facies, compared with much higher representation in the more organic sediments of floodplain pools of the stadials.

The graphs, Figures 8.2 to 8.10, show the representation of the following taxa, selected from categories 1a, 1b and 1c above, together with taxa of heath plants, added because of their significance for cold stage vegetation.

Category 1a

Gramineae (Figure 8.2) Devensian stadial. The l De and m De curves have a modal class in the 30 category and are positively skewed, with many samples having higher percentages. The e De curve is not so skewed, having a higher modal class, with a higher frequency of samples in the higher categories. The other curves (ml De, em De, De), with far fewer samples, show modal classes in the 20 to 40 range.

Devensian interstadials. The samples from the two e De interstadials (Br, Ch; sample totals very unequal) have contrasting curves. The e De Ch curve shows far lower Gramineae percentages than the e De Br curve, which has more affinity with De stadial curves. The UW (Upton Warren Interstadial Complex) curve shows a form similar to stadial curves.

Pre-Devensian cold stages. Curves from pre-Devensian cold stages show modal classes in the 20 to 40 range, with far fewer samples in the higher per-

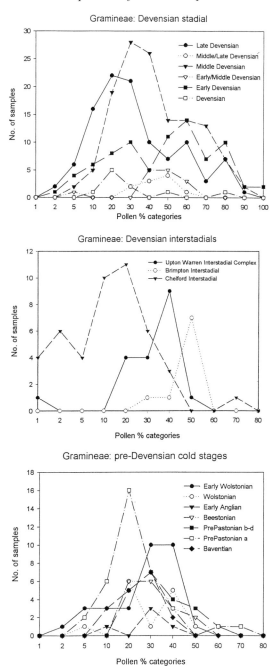

Figure 8.2. Cold stage representation of Gramineae pollen.

centage categories compared with the De stadial samples, which difference may be partly taphonomic in origin.

Cyperaceae (Figure 8.3) Devensian stadial. The De stadial curves show a wide range of the higher frequencies, more variable than the Gramineae curves, but again with few samples in the lowest categories.

Devensian interstadial. The e De Br and e De Ch curves are similar in form, with modal clases of 10 and 5 respectively. They contrast with the De stadial and UW curves, the latter showing more affinity with the stadial curves.

Pre-Devensian cold stages. As with the Gramineae curves for pre-Devensian cold stages, the Cyperaceae curves show lower modal classes (20–40), than the De stadial curves, with fewer samples in the higher percentage categories.

Category 1b

Pinus (Figure 8.4) Devensian stadial. The De stadial curves show a modal class of 5–10. Consistently low pollen percentages, with few samples rising to higher classes, appear characteristic of De stadial times. The presence of *Pinus* is not supported by the macro record, and the pollen is interpreted as having a long-distance origin. Samples with higher percentages appear more frequent in e De times, possibly a reflection of less distant coniferous forest in the Early Devensian.

Devensian interstadial. The e De Ch curve shows the expected higher frequencies of *Pinus* pollen which are the basis of definition of the interstadial. The e De Br curve, based on far fewer samples, is very different, with a modal class of 30. The UW curve resembles the stadial curves.

Pre-Devensian cold stages. These curves show modes in higher classes than the De stadial curves. The differences may result from local source areas, but they may also result from reworking of pollen from the preceding temperate stages. Taphonomic factors associated with marine sediments of the earlier cold stages may also play a part.

Cruciferae (Figure 8.5) Devensian stadial. The De stadial curves show a modal class of 1, except for the em De curve at 10–20. A few samples show higher frequencies, indicating more local abundance. The presence of Cruciferae throughout the periods represented is well supported by the macro record.

Devensian interstadials. A modal class of 1 is seen in the both e De Br and e De Ch interstadials, but there is no tail of higher frequencies, as with

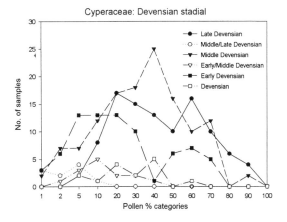

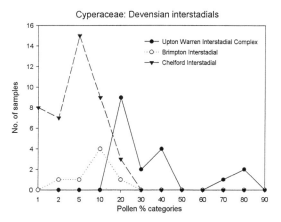

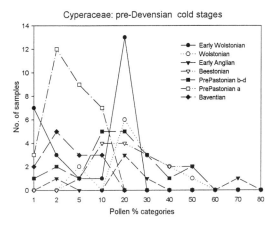

Figure 8.3. Cold stage representation of Cyperaceae pollen.

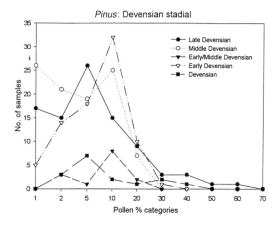

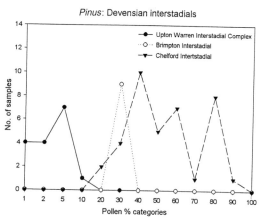

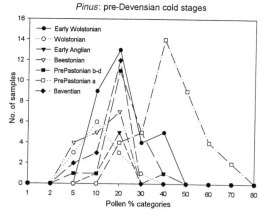

Figure 8.4. Cold stage representation of *Pinus* pollen.

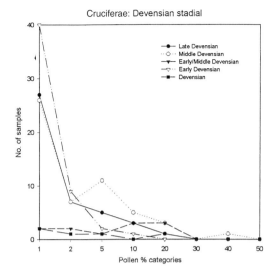

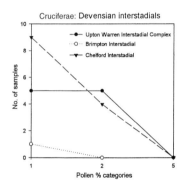

Figure 8.5. Cold stage representation of Cruciferae pollen.

the De stadial curves. This tallies with the macro record, which shows very rare occurrences in the Br and Ch interstadials while abundant Cruciferae macro remains occur in the De stadials and the UW interstadial. The difference must largely reflects the preservation of assemblages derived from Devensian stadial and UW interstadial floodplains as against the interstadial woodland facies of pool character.

Artemisia (Figure 8.6) Devensian stadial. The De stadial curves show modal classes of 1, with fewer samples showing higher frequencies, as with Cruciferae. But few macro remains of *Artemisia* are recorded in the Devensian. The achenes are less substantial than the seeds and fruits of

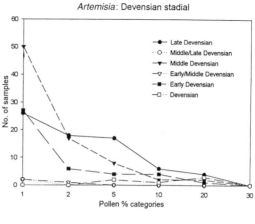

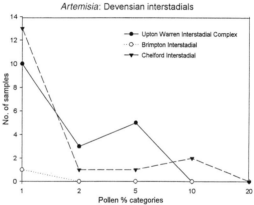

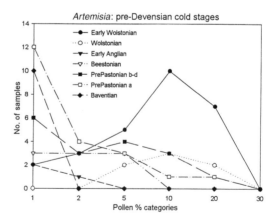

Figure 8.6. Cold stage representation of *Artemisia* pollen.

Cruciferae, but the pollen from this anemophilous genus is likely to be well dispersed, compared with the entomophilous Cruciferae, giving rise to the high number of samples (425) with *Artemisia*.

Devensian interstadial. The e De Ch interstadial curve has a similarity to the De stadial curves, suggesting a regional origin of the pollen associated with the interstadial conditions. The m De UW curve is also similar. The e De Br representation is considerably lower.

Pre-Devensian cold stages. With fewer samples, the curves are similar to the Devensian stadial curves, except that the e Wo curve shows a markedly higher representation.

Category 1c

Corylus (Figure 8.7) Devensian stadial. *Corylus* is interpreted as a reworked or distantly derived taxon in the cold stage flora. The De stadial curves show a modal class of 1, with few samples in higher categories. The e De and m De stadials show the highest number of samples in category 1, with the l De stadial having a much lower representation. The higher frequencies in the earlier part of the Devensian may relate to reworking of pollen from the previous temperate stage (Ipswichian) where the pollen of *Corylus* is abundant.

Devensian interstadials. The curves for the m De UW and e De Ch interstadials, based on far fewer samples, are similar to those of the De stadials.

Pre-Devensian cold stages. The curves of the pre-Devensian cold stages are similar to those above, the e Wo samples showing higher frequencies than the other periods, interpreted as the result of reworking of underlying temperate stage sediments.

Heath taxa

Graphs for Ericales, *Calluna* and *Empetrum* are illustrated to indicate the possible importance of heath taxa in cold stages.

Ericales (Figure 8.8) Devensian stadial. Low frequencies are recorded in the De stadials, except for e De, where higher frequencies are seen. These are mainly recorded from the Wing site (see Chapter 12).

Devensian interstadial. The m De UW and e De Ch interstadial curves show the same low representation as the l and m De curves.

Pre-Devensian cold stages. The pre-Devensian curves are divided into two categories. The Wo, An and Be curves show modes in the low categories, while the earlier periods, associated with marine sediments, show higher frequencies.

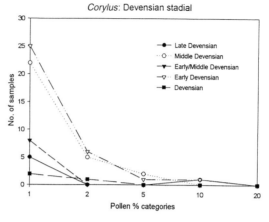

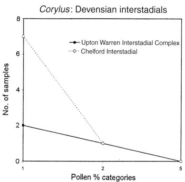

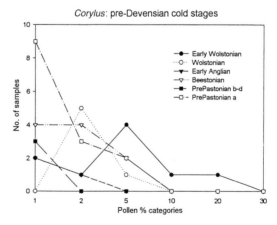

Figure 8.7. Cold stage representation of *Corylus* pollen.

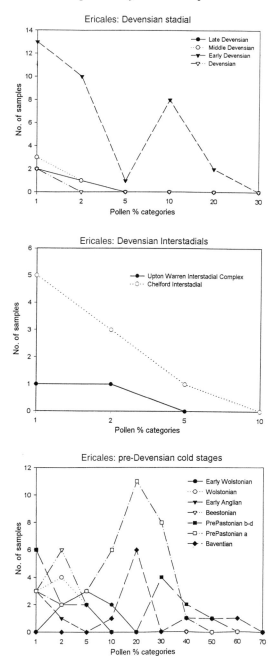

Figure 8.8. Cold stage representation of Ericales pollen.

Calluna (Figure 8.9) Devensian stadial. The De stadial curves resemble those for *Empetrum*, but there are more samples, with the e De the period with the highest number of samples with *Calluna*, as with Ericales.

Devensian interstadials. The e De Ch interstadial curve shows a larger number of samples with higher frequencies, associated with interstadial woodland assemblages. The m De UW curve resembles the later Devensian stadial curves.

Pre-Devensian cold stages. Pre-Devensian cold stages show a lower frequency of pollen than Ericales and *Empetrum*, with representation similar to the Devensian stadials.

Empetrum (Figure 8.10) Devensian stadial. In the De stadials modes are in the lowest category, with relatively few samples.

Devensian interstadial. In the only interstadial represented, e De Ch, the curve is similar to those of the Devensian stadials.

Pre-Devensian cold stages. As with Ericales, there are higher frequencies associated with the pre-Devensian cold stages, especially Pre Pa a.

These records for heath taxa indicate that apart from the e De Ch interstadial, the e De stadial and the earlier pre-Devensian cold stages, they made only a minor contribution to the pollen assemblages which have been preserved, though samples with a few grains occur more regularly. It appears that though heath communities are identifiably present at particular times, the low contribution of pollen at other times suggests they were notably rare.

Comments on representation

The details of macro and pollen representation outlined above offer an opportunity to summarise and discuss some general aspects of the interpretation of the fossil assemblages. Pollen analysis leads mostly to the identification of families and genera, with more possibilities of species identification from macro remains. The pollen data are significant for the recognition of local and regional vegetation and flora, while the macro data derives from more localised plant communities. There is thus more information on local vegetation and flora from macros, while pollen can provide more information on regional vegetation and flora but at a lower level of species recognition. Variation of macro assemblages can reveal variation of local plant communities, while variation of pollen assemblages may result from the same or from variation in regional vegetation and flora.

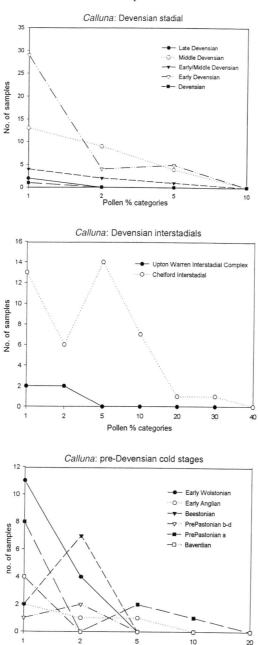

Figure 8.9. Cold stage representation of *Calluna* pollen.

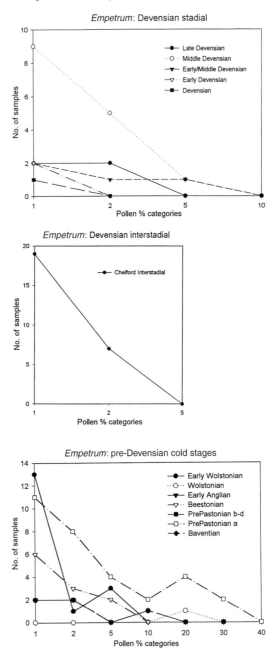

Figure 8.10. Cold stage representation of *Empetrum* pollen.

Overlying these relationships is the variation which may result from processes of taphonomy acting on the formation of the fossil assemblages.

Factors affecting the constitution of assemblages can therefore be grouped as follows.

1. Biota-dependent
   a. Vegetation-dependent. The structure of vegetation controls the constitution of pollen assemblages dispersed from vegetation, as is widely recognised in studies of the relation between forest composition and pollen rain. The same applies to herb vegetation, where height of flowering, time of flowering, pollen production and pollination methods will affect the dispersal of a pollen assemblage. The major pollen taxa in category 1a above, Gramineae and Cyperaceae, are anemophilous and usually flower at a height above other members of the local plant community (Plate 12.4B). Entomophilous taxa, which may be important in the plant community, are recorded at lower frequencies in the fossil assemblages, and they often flower at a lower height. Similar effects may be present with the dispersal of macroscopic remains, fruiting heads of herbaceous species often borne on stalks elongating as fruit sets, with dispersal by wind and perhaps later, water.
   b. Species-dependent. These are the properties of the species which determine production of potential fossils, e.g. pollen production, pollination method, seed, fruit or leaf production, characters conducive to preservation.
2. Taphonomy-dependent
   a. *Dispersal of assemblage*. The sedimentary environment determines the importance of taphonomy in altering the constitution of a dispersing assemblage (see Chapter 3).
   b. *Weathering* is another significant cause of change of the constitution of assemblages, either in course of transport or post-depositionally.
   c. The *sediment* containing the contemporary assemblage may contain reworked fossils from older sediments.

The relative importance of these factors will vary from site to site. Biota-dependent factors may be more important than taphonomy-dependent factors, or the latter may be more significant. The possibilities at a particular site must be considered in any interpretation of the fossil assemblages. This is particularly important in studies of cold stage floras, where there is much variation in sedimentary environments, many opportunities for reworking, and a complex of plant communities.

The consequence of these effects on assemblages is to blur changes in plant communities which might result from vegetation succession or climatic change. Differentiation of stadials versus interstadials is quite clear when the latter are forested, but less distinct changes of vegetation may result not only from climatic change, but, for example, from local plant succession (e.g. a hydrosere or a succession on a retrogressive thaw slope), changes in the sedimentary environment, changes in hydrology and permafrost distribution, and soil development. In addition to these problems of identifying change in time, it is even more difficult in the present state of knowledge to distinguish the regional variation of vegetation and flora which surely must have been present, allied to climatic gradients and the underlying geology.

9

Biological aspects of the cold stage flora

The nature of the cold stage flora may be more readily understood by examining some of its biological characteristics, and comparing them with those of arctic or more northern present-day floras. Many attributes of species of the present British Flora have been described, e.g. by Grime *et al.* (1990), and attributes of species of the cold stage flora could be analysed in a similar way. Thus, *Campanula rotundifolia* is described by these authors as a stress-tolerant species, which may assist in explaining its common occurrence in cold stage floras (Table 8.8). But knowledge of this kind is confined to only a part of the present flora, and knowledge of the fossil flora is also incomplete, so any detailed comparisons would be at present of limited application. However, some general aspects of the biology of the cold stage flora can be usefully discussed.

The PLANT table is a list of identified taxa, mostly at a species level, giving certain biological attributes of the taxa. This table is the basis for the following discussion of life form, variation of species (both taxonomic and ecological), and other aspects of the flora.

Life form

Raunkiaer (1934) developed a classification of plants based on how they are adapted to survive the unfavourable season. The classes, the life forms or biological types of Raunkiaer, are defined on the position of resting buds or persisting stem apices in relation to soil level, with an additional class of plants surviving the unfavourable season as seeds. The classification is shown in Table 4.9.

Raunkiaer also developed the idea of a biological spectrum of a flora of a particular climatic region, which showed the proportions of different life forms in that region. Such an analysis would show, for example, the

Table 9.1. *Life form analysis of the cold stage flora*

Class	No. of taxa ('a' or 'da' in PLANT table; no. in parentheses includes taxa with >1 life form)		Cold stage %	Arctic Ch %[a]	Denmark %[b]
MM 7 (7)					
M 2 (5)	Ph 20 (31)		6 (7)	3.5	7
N 11 (19)					
Ch 7 (12)					
Chw 4 (9)	Ch 36 (51)		12 (12)	19	3
Chh 22 (26)					
Chc 3 (4)					
H 20 (31)					
Hp 26 (35)	H 118 (155)		38 (35)	64.5	50
Hs 61 (75)					
Hr 11 (14)					
G – (–)					
Gb 1 (1)					
Gr 1 (1)	8 (15)				
Grh 5 (9)					
Grt – (3)					
Gt 1 (1)		Cr 88 (140)	29 (31)	10	22
Hel 31 (63)	80 (125)				
Hyd 49 (62)					
Th 45 (63)			15 (14)	3	18
Total 307 (440)					

[a] Arctic Chamaephyte biological spectrum of Raunkiaer (1934).
[b] Denmark biological spectrum of Raunkiaer (1934).

proportion of annuals in an arctic flora, or of trees in a lower latitude temperate flora, so relating a climate to a biological spectrum. He paid detailed attention to arctic floras studying particular areas and combining the results to define an Arctic chamaephyte climate. Such definitions will depend on the size of the area considered, size of the flora considered and the range of habitats covered.

Raunkiaer's system offers one way of looking at a fossil flora and comparing it with extant floras in a very general way, particularly when seasonal climates promoting over-wintering are concerned, as must be the case in cold stages. Table 9.1 lists the number of 'a' or 'da' taxa (species) in the PLANT table belonging to Raunkiaer's life forms, and shows the biological spectrum of the 'a' and 'da' taxa. A detailed comparison of this with Raunkiaer's biological spectra from different climatic regions is hardly justified, since the cold stage flora preserved fossil is only partially representative of the total flora, is moreover biased by particular habitats favouring preservation, and the flora list is from probably varying climatic conditions within cold stages.

However, with a total of over 300 'a' and 'da' taxa, there is an opportunity to prepare a cold stage flora biological spectrum, and to compare this with Raunkiaer's spectra from climatic regions. Table 9.1 compares the cold stage flora spectrum with those calculated by Raunkiaer for his Arctic Chamaephyte climate and the temperate region of Denmark. A comparison of these shows the following.

The Phanerophyte percentage of the cold stage flora is greater than the Arctic percentage, similar to the Denmark percentage.

The Chamaephyte percentage is higher than the Denmark percentage, approaching the Arctic percentage.

The Hemicryptophyte percentage is lower than the Arctic and Denmark pecentages; a high value was considered by Raunkiaer to be characteristic of temperate regions.

The Cryptophyte percentage is higher than the Arctic and Denmark percentages, inflated by the number of Helophytes and Hydrophytes. Raunkiaer considered Cryptophytes were characteristic of climates with a marked unfavourable season, such as steppe in the case of geophytes.

The Therophyte percentage is much higher than the Arctic percentage and approaches the Denmark percentage. Raunkiaer considered that therophytes were characteristic of regions with hot or dry climates, such as steppe or desert, where the unfavourable season was survived as seed, and that higher percentages were favoured by open ground conditions.

The comparison shows the distinctiveness of the cold stage flora spectrum, in particular the high percentage of therophytes. These are very rare in present Arctic floras. The growing season is too short, reproduction by seed is uncertain, and soil conditions with shallow permafrost and freeze/thaw movements are not favourable for establishment. At lower latitudes, with warmer and longer summers, therophytes may be more abundant in alpine floras. Bliss (1971) in a discussion of Arctic and Alpine

Table 9.2. *Life spans*

Life span (years)	No. of taxa ('a' & 'da' in PLANT table)
Annual (1)	44
Biennial (2)	9
Perennial (3)	293
Varied (1–3)	31
Biennial/Perennial (2–3)	9

annuals, points out that annuals in the Sierra Nevada of California relate also to those of the desert below, rather than being arctic-alpine in distribution. A similar explanation can be applied to the increased number of therophytes in the cold stage flora, able to take advantage of the longer summer of the lower latitude and of open ground conditions.

Life span

Table 9.2 lists the number of species in the PLANT table in terms of life span. The number of annuals is discussed in the previous section. The number of biennials is low. It is also low in the present arctic flora (where such species may behave as biennials in the higher Arctic, and annuals in the low Arctic where they can complete their annual cycle more easily). Many taxa have variable life spans, including the commonly recorded *Viola* and *Draba*. As with the Arctic, the majority of species are perennial. Some may be short-lived perennials, compared with long-lived temperate phanerophytes. Populations of short-lived perennials may provide a wider genetic base for adaptability to changing climate and environments.

Variation

Those engaged in identifying macroscopic plant remains from cold stage floras soon realise that many identifications are made difficult by the variability of species commonly found. The taxonomy of such species may be complex and problematic, and descriptions or reference material of variants is often inadequate. Polunin (1959) summed up the problem thus:

By a devilish combination of apomixis, hybridisation and morphological plasticity, the genus *Potentilla* is rendered one of the most difficult and generally 'critical' in the Arctic.

The presence of variable species in the cold stage flora is of great interest, perhaps related to the diversity of habitat discussed later. Although identification of a fossil to a subspecies or other variant within a species or species aggregate is mostly impossible at present, the variability of the living species is a guide to variability in the fossil species. Evidence of polymorphism or ecotypic variation must rest on comparisons with the modern species. Examples of present-day variation of phenotype classes within populations of cold stage species have been described by Vavrek *et al.* (1997) for *Taraxacum officinale* and by Sultan *et al.* (1998) for annual *Polygonum* species.

Variation in a species is expressed in many ways, as polymorphism, phenotypic variation and ecotypic variation, controlled by the reaction between the genotype and the environment. Variation seen in the modern species is likely to have been partially a result of environmental conditions during the last cold stage, acting on the variety of open habitats available in largely herbaceous vegetation, similar to the kinds of variation well known from the modern Arctic flora (see Crawford & Abbott (1994). The concept of 'hybrid zones' (Harrison 1993) considered in relation to plants in cold stages, with significant introgression, such as has been described for populations in disturbed habitats, would favour variation, with maintenance of genetic diversity (Reiseberg & Wendel 1993).

Such habitats may favour polyploid and apomictic plants (Briggs & Walters 1997). In this respect, though the data are very far from perfect, a tally of diploid chromosome numbers given in the PLANT table (from Clapham, Tutin & Moore (1989) and *Flora Europaea*) for the 'a' taxa (species) shows that *c.* 65 per cent have a single diploid number given, with the remainder polyploids or multiple numbers. The frequency of the latter is in apparent contrast to the higher frequency of polyploids found in the Arctic flora, perhaps a reflection of the less severe and possibly more stable conditions of the cold stages compared with the present Arctic.

Variable species

While some cold stage species of widespread occurrence show little variation (e.g. *Menyanthes trifoliata*), many are very variable (e.g. *Helianthemum canum*, *Campanula rotundifolia*, *Armeria maritima*). Table 9.3a shows the number of species recorded as variable, very variable and very polymorphic in the PLANT table, the information based on the species accounts of Clapham, Tutin & Warburg (1962) and *Flora Europaea*. The total, 104 species and three species aggregates, is about 30 per cent of the total of species identified (364), a high percentage which may reflect an important property of the cold stage flora. Another way of estimating the number of variable species in the flora is to list the number of subspecies identified. This analysis is shown in Table 9.3b, with the information again based on the floras mentioned above. The number reflects the complex taxonomy of a large number of species in the cold stage flora. The families with the highest numbers of species in this table are Caryophyllaceae (15), Compositae (15), Rosaceae (10), Ranunculaceae (7) and Cruciferae (7).

Phenotypic and physiological plasticity

Porsild's knowledge of the Arctic flora led him to observe that

Common for the widely distributed arctic species is their astounding tolerance and adaptability to large amplitudes in day-length, from the extreme of 24 hours of day-light throughout the growing season in Ellesmere Island to a maximum of 15 hours and 10 minutes in the latitude of Pike's Peak in Colorado, more than 43 degrees of latitude south of Cape Columbia in Ellesmere Island . . . Perhaps the answer is that several, if not all, of these species are composed of ecotypes having different photo-period requirements.   *(Porsild 1955)*

Earlier, Turesson's classical experiments demonstrated the presence of ecotypes in many species, with habitats correlated with genetic variation. For example, he demonstrated the existence of alpine and phenological ecotypes of *Campanula rotundifolia*, a common cold stage plant (Turesson 1925). Since then physiological plasticity, sometimes associated with partic-ular phenotypes, has been demonstrated in many species with an Arctic dis-tribution, for example *Oxyria digyna* (Mooney & Billings 1961), *Saxifraga oppositifolia* (Crawford 1997; Crawford *et al.* 1995) and *Draba* (Brochmann *et al.* 1992). Selection of ecotypes can occur over short periods (decades) (Bradshaw & McNeilly 1991), significant considering the short generation time of many cold stage taxa.

Such is the background for the observed variability of many species whose fossils have been identified in cold stage floras, not only those which are now considered Arctic in distribution, but also those which now have a more southern distribution.

Growth, reproduction, competition, dispersal

Salisbury (1942) in his study of the reproductive capacity of plants noted that vegetative multiplication was favoured at higher latitudes over seed production because physiological requirements for seed production

Table 9.3. *Numbers of variable species and
subspecies*

(a) *Variable species*

	Species	Species agg. or *s.l.*
Variability remarked (V)	38	–
Very variable (W)	58	2
Very polymorphic (P)	9	–

(b) *Subspecies*

No. of species	No. of subspecies recorded for the species
12	1
44	2
20	3
9	4
13	5
7	6
5	7
3	8
1	9
1	12
1	14
1	15
1	24
1	104

demanded higher temperatures than those required for growth. Vegetative multiplication and spread by low rooting branches, rhizomes, stolons and bulbils is indeed common in the Arctic and compensates for the uncertainty of seed production in a severe climate with a short and unreliable growing season, though the observed survival of seeds in frozen soils balances this.

Such conditions were not likely to have been the case in lower latitudes, where more stable and lengthy growing conditions prevailed and where the summer active layer might well be deeper than in the Arctic. These conditions would favour several aspects of plant life, including increased rooting depth, improved possibilities of pollination with a richer insect fauna (nearly 60 per cent of the 'a' and 'da' species in the PLANT table are entomophilous), of seed set and of the success of soil seed banks. It is notable that many of the species listed most commonly in north-west European soil seed banks by Thompson, Bakker & Bekker (1997) (mainly grassland, to a lesser extent arable and a diversity of habitats) are also well-recorded in the cold stage flora. Such species include *Urtica dioica, Polygonum aviculare, Rumex acetosella, R. acetosa, Stellaria media, S. graminea, Ranunculus repens, R. acris, Filipendula ulmaria, Potentilla erecta, Plantago major, P. lanceolata, Campanula rotundifolia, Achillea millefolium, Taraxacum officinale* and *Juncus bufonius*.

In these environments at our latitudes, competition would be more significant than in the higher Arctic, where control of plant distribution is more likely to result from climate conditions than competition (Young 1971), lower in the unproductive environment.

As in the Arctic, however, seed dispersal was probably favoured by cold stage conditions, with no barriers of tall vegetation to wind dispersal, with wind and frozen surfaces aiding dispersal. Interbreeding populations might then have been very extensive though localised.

Consequences

The discussion briefly indicates some ways in which the cold stage flora appears to differ from more northern present-day floras. These differences give rise to consequences for the interpretation of the cold stage flora and dependent biota in terms of the present behaviour of species and their distribution and ecology.

The mainly herbaceous vegetation of the cold stages is evident from the life form analysis. In contrast with forested periods, the shorter generation times of the plants and the varied open habitats give a very different aspect to evolutionary processes of cold stage herbaceous plants. The nature of the vegetation is discussed in a later chapter, but here it is suggested that there is little relation to present Arctic vegetation. At the latitude of the British Isles the vegetation is likely to have been more productive, with consequences for enriching the fauna.

It is also worth briefly noting some aspects of insect/plant relations in the present Arctic, from the point of view of comparison with cold stage conditions at lower latitudes. In the Arctic, Diptera are better represented in the fauna than Coleoptera, and pollination is by flies and bumble bees. At lower latitudes Coleoptera outnumber Diptera (see e.g. Danks 1990). In the Siberian tundra, there are more entomophilous plants than anthophilous insects, a situation which becomes reversed to the south (Chernov 1985). Nectar yields of entomophilous flowers tend to be lower in the High Arctic than the Low Arctic (Pielou 1994), with less specialised pollination mechanisms are present.

The question to be asked is how such characteristics of insect life in the Arctic and the relations to plants in the Arctic are changed in a cold stage environment at the latitude of the British Isles, with very different seasonal conditions. Did the summer conditions lead to a richer Coleopteran fauna than the restricted fauna of the far north? Was seed set of entomophilous plants greatly enhanced by a richer insect fauna? What is the consequence of the observation by Aurivillius (1883) that in the Arctic

those insects which as larvae collect their food from living parts of plants, become fewer to the north or else disappear completely, whereas on the other hand those which live in water as larvae or else among rotting plant material, plus some of those which are predators, survive best

if such larvae are able to collect plant food in the richer vegetation of the cold stages. Arctic species are clearly adapted to rigorous conditions: but how does their distribution and behaviour change at the lower latitudes, and how does this affect cold stage plant life and insect communities?

10
Habitats of the cold stages

Habitat and vegetation type are closely woven. But before considering the latter in more detail, a survey of the evidence for cold stage habitats will usefully set a background for analysis of the vegetation. At the same time, the ways in which habitats may be controlled by environmental conditions can be brought into the discussion. A strong contrast will be apparent here between the great variety of cold stage habitats and the forested landscapes of the temperate stages, with microhabitats greatly muted by the forest canopy.

It is essential to understand the habitat variety in cold stage environments (e.g. Plates 4, 5, 6, 7A), especially since it makes for great difficulties in interpreting the make-up of the vegetation. Variety can be on a micro scale and on a wider regional scale; both can reasonably be assumed. Evidence for the variety of habitats comes from a number of sources, including the present landscape, geology, sediments and palaeontology.

Landscapes
During the last cold stage, outside the area covered by ice during ice advances (Figure 2.2), the landscapes in the periglacial area must have been generally much the same as they are at present, with plateaus, scarps and valleys controlled by the geology seen today. The variety of rocks will have affected soils, and so habitats and vegetation, probably more so than today, with active periglacial processes renewing soil-forming processes. The same can be said about the periglacial areas in the south of the British Isles outside the limit of maximum glaciation during the older cold stages.

Since most cold stage sites are in the south and east of Britain, most is known of conditions in this area, which lies south and east of Tansley's 'Highland Line' (1939) (Figure 2.2). There are only few sites in the hill and mountain regions of harder rock and more relief to the west and north. In morainic areas of fresher and younger landscapes vacated by ice but not degraded by periglacial processes, variety of habitat is likely to have been increased by greater variety of aspect and so insolation. Aspect can be all-important in determining vegetation, as, for example, demonstrated by Walker *et al.* (1991) in their account of steppe vegetation on the south-facing slopes of pingos on the Alaska Coastal Plain.

Solifluction, cryoturbation and patterned ground (Figures 10.1, 12.1) must have produced a great variety of habitats in the upland areas, with gravel floodplains and meander pools in the valleys. To this variety must be added the variety of parent rock types, e.g. in the south-east area, a range from porous sandstones to chalk to clay.

In summary, the main factors controlling the nature of the open cold stage habitats are: parent rock lithology, permafrost susceptibility of the parent rock and its soil, the presence of solifluction and patterned ground, drainage conditions and the position of the water table, dependent on permafrost or parent rock lithology, and aspect. Drainage can be critical in, for example, polar desert, where impedance may lead to the only closed herbaceous communities, such as sedge fields, in a region (Figure 12.1, Plate 3B). Active solifluction may favour growth of open herb as against closed herb communities on more stable slopes (plate 5A).

Soils

A brief survey of the variety of soils which might be available for cold stage plant communities is necessary to help understanding of the habitats and vegetation. Again there is the problem of finding modern analogues which might parallel soil-forming processes in cold stages at our latitude. Soil types of present northern areas, with generally low soil temperatures and low nutrient availability, may give significant indications of the kinds of soil present in cold stages, but the latitude differences, with implications for seasons and insolation, must have been substantial.

A suitable area for comparison has to be one with relatively soft rocks and also one in which terrestrial peat-forming processes are not prevalent. Such processes are nowadays very widespread in the north, supporting the presence of organic soils and the growth of shrub tundra, usually over permafrost. But there is little evidence for them in the cold stage record of sediments and vegetation. Conditions at more southern latitudes do not promote widespread growth of organic sediments, apart from local accumulations of organic sediments associated with impeded drainage. With lack of peat and litter, summer warmth penetrates the soil more rapidly, promoting a deeper active layer if permafrost is present. Thus there appears to be a major difference between certain present northern soil conditions and cold stage conditions, which should be kept in mind in considering cold stage environments and plant (and animal) distributions.

Tedrow (1973) made a broad classification of soils of northern North America into polar desert soils, sub-polar desert soils and tundra soils. Since then, much more detail of northern soils has been acquired and described, e.g. in the Canadian system of soil classification (Agriculture Canada Expert Committee on Soil Survey, 1987).

Such description is available for a useful and exemplary analogue area, Herschel Island, north of the shrub tundra region of Canada, off the Yukon coast, at 69° N. The island is formed of morainic deposits of the Last (Wisconsin) glaciation, with mainly mineral soils and no great peat formation, and is underlain by permafrost. The plant communities show a close relation to soils, controlled by slope, drainage and depth to permafrost (Smith *et al.* 1989). The major soil group, using the Canadian system of soil classification, is cryosolic, with a shallow active layer over permafrost. The cryosolic soils may be turbic (affected by cryoturbation), or static and regosolic where there is more active deposition of sediment or where mass movement has hindered soil development. Brunosolic soils are present on the more mature and level landscapes, where soil development has proceeded. All these soils have their typical mainly herbaceous vegetation, giving rise to a complex mosaic.

Such mosaics must have existed in cold stages at times of permafrost, especially in areas of patterned ground (Figure 10.1; Watt *et al.* 1966). But at our latitude, active layer depth is likely at times to have been greater, encouraging a greater range of soils. Where the active layer is thick, as it may be in gravelly and sandy river floodplains, the regosol may develop more warmth than in wet tundra conditions, giving rise to much more favourable conditions for germination and plant growth. At our latitude such differences will be magnified. The environments of sandy and gravelly floodplains would then provide a variety of favourable habitats, reflected in the rich cold stage herb floras in their sediments, a contrast with less well drained upland plateau areas.

The presence of wide and narrow thermal contraction cracks in cold stage sediments (see e.g. West 1993) indicates variation in drainage and dryness of various habitats. In the High Arctic today, drainage is a major factor, with well-drained regosols promoting polar desert and impeded drainage sedge fields.

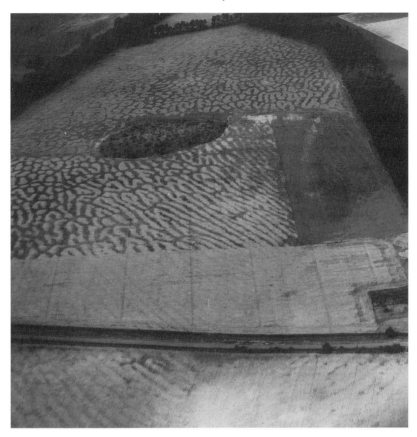

Figure 10.1. Patterned ground at Brettenham Heath, Thetford, Norfolk, origin in the Devensian Stage, now manifested as vegetation patterns. Networks are seen on level ground, stripes on the shallow valley slopes in the right hand bottom quarter. Photo by J.K. St Joseph; Crown Copyright Reserved, RAF Photo.

Soil disturbance by freeze/thaw and solifluction lead to favourable conditions for species diversity in arctic and alpine tundras (Fox 1981), with variability of soil conditions and plant succession. Such will also have been the situation in cold stages, again perhaps magnified by the lower latitude.

Salinity

The occurrence in cold stage floras of plants which are nowadays halophytes has long been known. Their occurrence has been discussed by Bell (1969, 1970), who related their presence to saline soils, such as are developed in an arid continental climate, probably developed over permafrost. Alternatively, Adam (1977), discussing the ecology of halophytic species (both 'obligate' and 'facultative'), proposed that ecological factors such as reduction of competition in the open cold stage landscape allowed the spread to inland habitats of species nowadays found in coastal salt marshes.

It seems probable that both these explanations contribute to the presence of halophytes in old stage floras. Halophytes are widespread in inland regions with a strong continental climate in northern Eurasia (see e.g. Cajander 1903; Printz 1921) and northern America (e.g. Brandon 1965), indicating their ease of migration inland (Plate 7A). In such places, evapo-ration leads to deposition of salts, particularly in areas of permafrost, where water draining from summer active layer melting can concentrate in low depressions, with the production of brackish water and saline soils. Groundwater discharge is often associated with such conditions. In polar desert areas, salt crusts and efflorescences develop on the surface, through evaporation and rise of capillary waters from fine-grained sediments below (French 1996). In the Late Devensian at the Fenland site of Somersham, troughs of silt at the top of massive thermal contraction cracks showed des-iccation cracks and horizons filled with calcite, evidence of evaporation and arid conditions (West 1993).

A further possible source of salinity is in waters derived from rocks. For example, in East Anglia at present, water from boulder clay and connate water of the limestones or sands in certain Jurassic clays are saline, enough so to be above the potable limit (Woodland 1946). Today, pools with a brackish fauna and flora occur in deep brick pits in Oxford Clay near Peterborough (Dogsthorpe Star Pit Site of Special Scientific Interest) as a result of the geology.

Taking into account the observations on present-day saline soils and the inland distribution of halophytes, the nature of the cold stage seasonal cli-mates, the presence of permafrost and possible groundwater conditions, it seems hardly surprising that halophytes are found in cold stage floras.

Sediments

Modification of soil conditions can occur through deposition of cover sand and loess. Both these will modify surface conditions and tend to produce a reduction in any variability of soil already developed. Cover sand deposi-tion will allow better surface percolation, drier and therefore warmer soils in summer, though wind blast may localise the development of vegetation

(Plates 4, 7B). Loess deposition has been shown to increase soil fertility indices, with increased biomass and species diversity, in an area of the Yukon receiving loess from the Slims River delta (Laxton *et al.* 1996), a matter relevant to the grazing population.

The scale of the processes leading to the variety of habitats considered above can itself vary in area and in time. For example, cover sand deposition may be local, coeval with many other states of soil development and drainage. The problem of segregating plant communities via fossil remains in such conditions will be apparent.

Habitats

A way of surveying habitats available for cold stage plants is to analyse the fossil taxa with the purpose of listing the habitats they normally occupy at present. The data in the PLANTHAB table, compiled from Clapham, Tutin & Warburg (1962) and the *Flora Europaea*, gives in simple form the pre-ferred habitats and certain other environmental factors for taxa where the level of identification is sufficient to do so. Table 10.1 gives the total number of taxa in the 'a', 'm' and 'da' use categories in the fields of the PLANTHAB table. Separate figures are given for taxa with a normal preference for a single habitat type and for taxa with a wider preference. The figures thus give a measure of variability of habitats, as listed in the PLANTHAB table, occupied by the taxa.

The analysis makes a simple (or coarse) separation of vegetation catego-ries and place categories. Overlap here is inevitable, but the separation pro-vides different ways of summarising the variability. Other aspects of plant behaviour or environment are also in the analysis.

The number of fossil records for each taxon is not taken into account in the analysis, though the information is available in the database tables. The aim is to provide an insight into the variability of habitats and the number of taxa associated with them. The analysis in Table 10.1 certainly demon-strates an unexpectedly wide representation of habitats by the taxa found fossil, now summarised as follows.

Vegetation. The G category (grassland, downs) contains by far the largest number of taxa, followed by A (aquatic) and Ma (marsh) taxa. Other terrestrial taxa (Fl, D, Mo categories) are well represented. The W category (woodland, scrub) will include tree taxa found in intersta-dials. The number of taxa associated with heath or other acidic habi-tats (categories B, H, Mo) is low.

Table 10.1. *Number of taxa in habitat categories of* PLANTHAB *table*
(uses: 'a', 'm', 'da')

Habitat	Total	No. of taxa single preference	wider preference
Vegetation			
A aquatic	73	62	11
B bogs	21	2	19
D dunes	19	4	15
Fe fens, flushes	49	9	40
Fl fell, rocks, screes, ledges, crevices, barrens	57	33	24
G grassland,downs	111	64	47
H heath	34	1	33
Ma marshes	58	15	43
Me meadows	22	5	17
Mo moors	21	4	17
W woodland, scrub	49	19	30
Total	514	218	296
Places			
B bare ground, mud	17	3	14
Co coastal	34	6	28
Cg cultivated ground	52	5	47
Ds disturbed ground	3	–	3
Fi fields	5	–	5
Fl river/stream sides, banks lake edges	65	33	32
Gr grassy places, pastures	110	28	82
Gv gravelly places	27	3	24
H hedges	18	–	18
M mountains	60	23	37
P peaty places, muddy	26	17	9
Ro road/way sides	32	–	32
Rp rocky places, screes, cliffs	63	11	52
Sa sandy places	37	4	33
Sc scrub, bushy	10	–	10
Wa waste places	44	3	41
Wo woody places	24	3	21
Total	626	139	487
Water			
Da damp	103	88	15
Dr dry	38	37	1
Ds seasonal damp	–	–	–
We wet	55	37	18

Table 10.1 (*cont.*)

Habitat		Total	No. of taxa single preference	No. of taxa wider preference
Sh	shallow	67	24	43
Sc	shallow, calcareous	2	1	1
Wa	deeper, lakes	36	5	31
Wc	ditto, calcareous	4	3	1
Wo	ditto, oligotrophic	3	3	–
Total		308	198	110
Light (not aquatics)				
O	open	280	280	–
Sh	shady	18	18	–
Total		298	298	–
Salinity				
B	brackish	10	3 (6)	1
M	maritime	31	18 (5)	8
S	saline	8	–	8
Total		49	21 (11)	17
Soil				
A	acid, siliceous, base-poor	25	23	2
C	calcareous, basic, eutrophic	83	79	4
F	calcifuge, non-calcareous	14	12	2
H	heavy, clay	6	5	1
I	inorganic substrate (silt)	6	6	–
L	light, poor	7	4	3
M	mull	1	1	–
O	organic	3	1	2
S	saline	2	2	–
Total		147	133	14
Behaviour				
C	casual	7	7	–
R	ruderal	2	1	1
W	weed	10	9	1
Total		19	17	2

Place. The Gr category (grassy places, pasture) contains the largest number of taxa, as does the similar category in the vegetation analysis. The Fl category (river/stream sides, etc.) is next largest, with the terrestrial categories (M, mountains; Rp, rocky places, etc.) also being well represented. The number of taxa found in cultivated ground (Cg) and waste places (Wa) is large, indicating the importance of such habitats in cold stage conditions.

Water. The analysis here shows the large number of taxa associated with damp to aquatic habitats, not surprising since these have conditions most likely to lead to preservation of fossils. Plants of drier habitats are scarcer.

Light. The division is between light-demanding taxa and those favouring shade. The great majority are in the former category.

Salinity. A number of taxa are now found in maritime areas, fewer in brackish and saline habitats.

Soil. Soil requirements are various, but the greatest number of taxa favour calcareous soils, with fewer taxa which favour acid soils or are calcifuges.

Behaviour. A small number of taxa are described as casuals, ruderals or weeds. These categories will overlap with certain of the categories in the vegetation and places categories.

Table 10.1 compares the number of taxa with single habitat preference against the number with less strict requirements. From the habitat point of view, in the 17 categories of the places field the majority of taxa have wider preferences, as do those of the vegetation field, indicating the greater relative significance of taxa with wider preferences. But in the more specialised environmental fields of water, light, salinity and soil, single preference taxa are usually in the majority, indicating variation of habitats available within the broader vegetation and place categories.

11
The distribution of taxa found fossil

Distribution of the cold stage taxa can be considered either in space or in time. In the former, the present distributions of taxa found fossil can be examined to compare present with cold stage distributions. In the latter the record of cold stage taxa over several cold stages and in temperate stages of the Quaternary can be reviewed. Distribution maps showing the finds of species are not given, since there is much bias in the distribution of the sites. Such maps of certain cold stage species were published by Godwin (1975).

Present distribution of taxa

The present distribution of plants found fossil has often been used to assist definition of climates at the time they lived. As a preliminary to discussing cold stage climates and vegetation it is clearly necessary to examine present distributions and to consider their bearing on the interpretation of past climates and vegetation.

The present distribution of a species is a result of a complex of factors, both internal and external to the species. The internal factors involve the genotype, and its ability to react successfully to variation of external conditions, for example, by biotypic or ecotypic variations. Breeding systems and 'physiological plasticity' (Chapter 9) are examples of important internal factors.

The external factors acting on the species are equally complex. They include present climate, competition and historical factors which led to the present distribution, such as climatic history, barriers to or presence of migration routes, sea level change, and in recent time anthropogenic causes. Conditions particularly significant for cold stage plants and favourable to wide distribution of species are the treelessness, open conditions, low sea levels (up to *c*. 100 m lower than at present), migration corridors of large north- and west-flowing rivers in Europe, and the facility for dispersal of propagules in cold winters with frozen ground and snow, as occurs in the Arctic at present. In some respects, recent anthropogenic factors reproduce some of these conditions.

The modern distribution then presents a single frame in a scene mobile in time. It is a summed result of the interaction of the internal and external factors over time, and a simple direct relation between present climate and distribution, assuming also an equilibrium between the two, can only be a first approximation to the relation between distribution and climate. This is especially so where little is known of the biotypic variation within a species. Taxonomically invisible characters may determine, to an unknown extent, the distribution of a species at present or in the past. Perhaps this is more true of short-lived herbs than long-lived trees, making it an important point when considering cold stage floras.

The PLANTDS1 and PLANTDS2 tables give various distribution categories of cold stage taxa identified to a level sufficient for analysis (species, 'a' and 'da' categories). Fields of these tables relate to a particular category or aspect of distribution (listed in Chapter 4).

The present distribution of the species indicated in the PLANTDS tables can be viewed in two related ways. One is to use the phytogeographical groupings of British species which have been developed over the years as a basis for analysing changes of distribution of species of the cold stage flora. The other is to examine the present distribution of individual cold stage species in terms of their present distribution based on dot-maps or other geographical or ecological information.

Phytogeographical groupings

Preston & Hill (1997) noted that

the major purpose of defining floristic elements is to establish categories with similar climatic requirements and which may have at least some similarities of dispersal history.

A comparison of the phytogeographical groups discerned in the present British flora with grouping of the species of the cold stage flora can throw important light on the origin of the flora and on the response of particular species to environmental change.

The grouping of species in the British flora on a geographical basis has a long history. Forbes (1846) described five distribution types (floristic elements). He related these to their origin in geological time. For example, the

Scandinavian or boreal type was regarded as survival from a former time when such boreal floras were more widespread in the epoch of the Northern Drift (Glacial Epoch). Preece (1995) has traced the history of this approach to phytogeography (and zoogeography), an approach which has been continued to the present day. Definition of floristic elements and their relation to geological events has been developed by Matthews (1937, 1955), Godwin (1975), Birks (1973), Birks & Deacon (1973) and, most recently, Preston & Hill (1997).

These groupings can now be discussed in respect of cold stage floras, but it is important to point out that though groupings can be assembled, they describe only the present geography of species, and do not mean that species of the same group have the same history. This is parallel to the view that present-day plant communities do not necessarily have great antiquity. Each species responds to geological changes according to its own constitution and external factors affecting it.

Matthews' analysis of distribution (*PLANTDS1*)

The classic division of species of the British flora into phytogeographical elements defined according to the areas the species occupy in Europe was made by Matthews (1937), following the lines set by Salisbury (1932) and earlier botanists. Matthews' scheme excluded species widespread in Europe and further afield, a group he reported as accounting for about 55 per cent (*c*. 815 species) of the British flora.

Table 11.1a lists the number of species placed by Matthews in the phytogeographical elements and the number of cold stage taxa in each of his elements. It enables a rough comparison between the two, rough because of the problems of allocating species to particular elements and because of the incompleteness of the fossil record. However, the comparison shows the variety of floristic elements in the cold stage flora. It also suggests a decrease in the oceanic elements, an increase in the northern elements, including the continental northern and the elements (rather wide-ranging) described by Matthews as Arctic-Subarctic and Arctic-Alpine. In the latter group Matthews, following Kulczynski (1923), distinguishes 'Historical Northern' species from 'Historical Tertiary' species, the former of historically northern origin migrating south, the latter of middle European origin migrating north on the retreat of the ice. This distinction is a very interesting one, but, as Matthews remarks, much more needs to be known of the taxonomy and geography of the species concerned to develop the concept.

Table 11.1. *Phytogeographical analyses of the cold stage flora*

(a) *Numbers of species in Matthews' phytogeographical elements compared with number in cold stages*

Phytogeographical element	No. of species	%	No. of cold stage species (single records)	%
Arctic-Alpine	76	11	32 (9)	34
'Historical Northern' species	26	4	12 (2)	12
'Historical Tertiary' species	50	7	20 (7)	21
Alpine	9	1	3 (2)	3
Arctic-Subarctic	30	4	8 (5)	8
Northern-Montane	25	3	4 (2)	4
Mediterranean	38	5	1 (1)	1
Continental Southern	127	19	10 (5)	10
Continental	82	12	7 (3)	7
Continental Northern	91	14	17 (5)	18
Oceanic Southern	74	11	3 (2)	3
Oceanic Western European	76	11	2	2
Oceanic Northern	19	2	6 (2)	6
Total	647		93	

(b) *Phytogeographical analyses of Godwin (1975) and Birks & Deacon (1973), Middle Weichselian, Midlands and East Anglia*

Phytogeographical element	% no. of species	Aggregate groups	% no. of species
1. Arctic-alpine	14.4	Arctic-alpine	19.5
2. Arctic-subarctic	3.2	1,2,3	
3. Alpine	1.4		
4. Northern-montane	5.4	Northern	17.6
5. Continental-northern	12.2	4,5	
6. Continental-southern	7.7	Continental	25.0
7. Continental-widespread	4.5	5,6,7	
8. Subatlantic-northern	4.5		
9. Subatlantic-southern	2.7	Southern	11.4
10. Subatlantic-widespread	6.8	6,9,12	
11. Atlantic-northern	0.5	Atlantic,	19.0
12. Atlantic-southern	0.9	Sub-Atlantic	
13. Atlantic-widespread	3.2	8,9,10,11,12,13	
14. Widespread	33		
No. of taxa	335		

Table 11.1 (*cont.*)

(c) *Numbers of species in the floristic elements of Preston & Hill (1997), compared with numbers in cold stages*
(first digit: major biome category; second digit: eastern limit category; Eurosiberian: eastern limit at 60°–120° E; Eurasian: eastern limit across Asia to a limit east of 120° E)

Phytogeographical element	No. of British & Irish native species	%	No. of cold stage species (single records)	%
1. Arctic-montane				
13 European	29		12 (6)	
14 Eurosiberian	6		1 (1)	
15 Eurasian	3		1 –	
16 Circumpolar	41		15 (3)	
Total	79	5.4	29	8.5
2. Boreo-arctic montane				
21 Oceanic	1		– –	
23 European	10		3 (1)	
24 Eurosiberian	2		2 –	
26 Circumpolar	25		12 (3)	
Total	38	2.3	17	5.0
3. Wide-boreal				
34 Eurosiberian	1		1 –	
35 Eurasian	1		1 –	
36 Circumboreal	17		8 (2)	
Total	19	1.4	10	2.9
4. Boreal-montane				
41 Oceanic	7		2 (2)	
42 Suboceanic	5		1 (1)	
43 European	27		3 (1)	
44 Eurosiberian	9		4 (1)	
45 Eurasian	5		2 –	
46 Circumpolar	50		13 (4)	
Total	103	7.0	25	7.3
5. Boreo-temperate				
51 Oceanic	8		– –	
52 Suboceanic	8		3 (1)	
53 European	48		14 (3)	
54 Eurosiberian	67		23 (6)	
55 Eurasian	38		15 (3)	
56 Circumpolar	64		27 (3)	
Total	233	15.8	82	24.1

Distribution of fossil taxa

Table 11.1(c) (*cont.*)

Phytogeographical element	No. of British & Irish native species	%	No. of cold stage species (single records)	%
6. Wide-temperate				
61 Oceanic	1		– –	
63 European	3		1 (1)	
64 Eurosiberian	11		6 (1)	
65 Eurasian	5		4 –	
66 Circumpolar	14		8 (1)	
Total	34	2.2	19	5.6
7. Temperate				
71 Oceanic	48		1 (1)	
72 Suboceanic	28		8 (4)	
73 European	297		43 (16)	
74 Eurosiberian	120		32 (11)	
75 Eurasian	38		12 (5)	
76 Circumpolar	26		8 (2)	
Total	557	37.6	104	30.6
8. Southern-temperate				
81 Oceanic	25		– –	
82 Suboceanic	54		5 (2)	
83 European	106		16 (8)	
84 Eurosiberian	81		17 (8)	
85 Eurasian	17		6 (2)	
86 Circumpolar	13		6 –	
Total	296	20.0	50	14.7
9. Mediterranean				
91 Mediterranean-Atlantic	69		2 (1)	
92 Submediterranean-Subatlantic	47		1 (1)	
93 Mediterranean-montane	6		1 –	
Total	122	8.3	4	1.2
Grand total	1481		340	
Eastern limit categories				
Oceanic	159	10.9	5 (4)	1.5
Suboceanic	142	9.6	18 (9)	5.3
European	526	35.3	93 (35)	27.4
Eurosiberian	297	20.0	86 (27)	25.3
Eurasian	107	7.2	41 (10)	12.1
Circumpolar	250	17.0	97 (15)	

Analyses of distribution by Godwin (1975) and Birks & Deacon (1973)

Table 11.1b shows the percentages given by Godwin (1975) of the major phytogeographical groupings in the Middle Weichselian (Devensian, as then defined, 45,000–15,000 BP) of the Midlands and East Anglia, and also the breakdown of these major groupings according to the more detailed categories of Birks & Deacon (1973), as given by Godwin (1975). These analyses support the diversity of the cold stage flora already demonstrated. The combined continental element is best represented, with the Arctic-alpine and Continental northern categories having high representation. The more detailed divisions of Birks and Deacon enlarge the picture of diversity, with representation of many minor floristic elements in the Atlantic and Subatlantic categories.

Certain other phytogeographical information is given in the PLANTDS1 table, such as Godwin's (1975) list of Late Weichselian (late-glacial s.l.) species with a restricted northern range in Scandinavia; 24 of the 31 listed are also present in the cold stage flora.

Preston & Hill's (1997) analysis of floristic elements (PLANTDS2)

This is a most detailed, and highly useful, analysis of the British and Irish flora, developing Matthews' approach, using categories of range in the northern hemisphere and taking more account of the widespread species. Species are placed in floristic elements based first on major biome categories and secondly on limits of distribution eastwards to Eurasia and circumpolarity. The categories are shown in Table 11.1c, with species numbers in the elements of the present British and Irish flora and in the cold stage flora.

A comparison of the percentage representation of categories in the present and fossil floras (Table 11.1c) shows higher percentages in the latter of the northern and wide-temperate elements (1–6), and lower percentages in the temperate and southern categories (7–9). In terms of the eastern limit categories, there is in the cold stage flora a reduction in the oceanic and suboceanic categories, a reduction in the European category, associated with the fact that most trees and shrubs are in the European temperate element, and an increase in the Eurosiberian, Eurasian and especially the circumpolar elements. The last difference is not surprising since floras of higher latitudes generally have poor regional differentiation.

In the most southern category, it is notable that the cold stage flora includes the often-recorded *Helianthemum canum* of the Mediterranean-montane element, and rarer records of *Damasonium alisma* of the Mediterranean-Atlantic element. The presence of these species in the cold stage flora shows the possibility of species occurring in present and past conditions which must have been very different.

The number of species of the vegetation and habitat categories of the PLANTHAB table in the different Preston & Hill (1997) categories is discussed in Chapter 12 (see also Table 12.1, p. 201).

Bell's (1969) elements of the cold stage flora

Bell's (1969) study of cold stage floras brought together for the first time lists of species from several Devensian sites, many radiocarbon-dated. Though not strictly all phytogeographical, Bell's three elements in the flora, southern, steppe and halophyte elements, can be included here. The species concerned are listed in Table 11.2. Bell discussed in detail the significance of the three groups of species in terms of environment. In the case of the southern species, associated with northern species in the mix already mentioned, possible explanations in terms of climate were discussed, and the presence of significant factors other than climate were noted. The complexities of interpreting climate from present distributions were underlined. The steppe species were considered to suggest a physiognomically steppe-like vegetation with dry mineral soil, not floristically a steppe, while the halophytes indicated saline soils.

Distribution according to the Ecological Flora Database (PLANTDS1)

Distributions of British higher plants have been analysed by Fitter & Peat (1994) for the *Ecological Flora Database (EFD)* to give and East/West Index and a South/North Index for species. Only a few species have not been given such indices. The indices are derived from the proportion of 10 km grid squares in which a species is recorded in each of 13 regions in Britain. Each region is weighted according to how far west or north it is, and the values are scaled to range from east to west or south to north by setting the species with the most easterly or southerly distribution to a value of 0 and the most westerly or northerly to a value of 10. Figures 11.1 and 11.2 show the percentage of species in each 0.5 index class for the E/W and S/N species indices. The figures show the proportions of tolerant or widely distributed and more narrowly distributed species.

A similar analysis can be made of the cold stage species ('a' and 'da' taxa in the PLANTDS1 table). The total number of cold stage taxa used is of course far smaller than the present flora totals. Figures 11.1 and 11.2 give the percentage of species in each class. Two curves are given for the cold stage flora, one the result of using all taxa, the other excluding all taxa recorded only once. No great difference in pattern emerges after this exclusion.

Table 11.2. *Bell's (1969) elements of the cold stage flora*

Southern species
   [a]*Corylus avellana* L.
   [a]*Carpinus betulus* L.
   [a]*Sambucus nigra* L.
   Ajuga reptans L.
   Aphanes arvensis L.
   Carex arenaria L.
   Corispermum sp.
   Damasonium alisma Mill.
   Diplotaxis tenuifolia (L.) DC.
   Groenlandia densa (L.) Fourr.
   Helianthemum canum (L.) Baumg.
   Lycopus europaeus L.
   Najas flexilis (Willd.) Rostk. & Schmidt
   N. marina L.
   Onobrychis viciifolia Scop.
   Potamogeton acutifolius Link
   Potamogeton crispus L.
   Potentilla fruticosa L.
   Ranunculus sardous Crantz
   Rumex maritimus L.
   Salix viminalis L.

Steppe element
   Linum perenne agg.
   Androsace septentrionalis L.
   Helianthemum canum (L.) Baumg.
   Artemisia spp.
   Blysmus rufus (Huds.) Link
   Festuca rubra L.
   Corispermum sp.
   Gramineae

Halophytes

Obligate:	Facultative:
Glaux maritima L.	*Armeria maritima* (Mill.) Willd.
Juncus gerardii Lois	*Atriplex hastata*-type
Suaeda maritima (L.) Dum.	*Blysmus rufus* (Huds.) Link
Triglochin maritima L.	*Carex arenaria* L.
	Carex maritima Gunn.
	Eleocharis uniglumis (Link) Schult.
	Festuca rubra L.
	Najas marina L.
	Plantago maritima L.
	Potamogeton filiformis Pers.
	Potentilla anserina L.
	Ranunculus sardous Crantz
	Ranunculus sceleratus L.
	Zannichellia palustris L.

[a] Considered reworked.

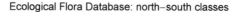

Ecological Flora Database: north–south classes

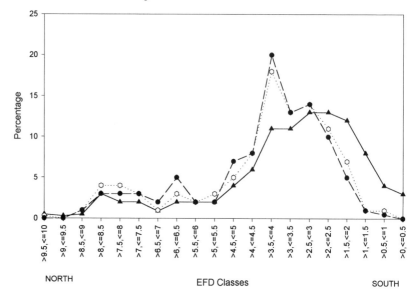

NORTH   EFD Classes   SOUTH

—▲— % EFD classes present British Flora (1598 spp.)
···○··· % EFD classes of cold stage species (318 spp.)
—●— % EFD classes of cold stage species with >1 record (221 spp.)

Figure 11.1. Analysis of the north–south distribution of the distribution classes of the Ecological Flora Database.

The S/N indices (Figure 11.1)

The curve for the present flora shows positive skew towards the southerly indices, with modal classes at 1.5–3. Lower numbers, the more northerly species, are in the classes above 5. The cold stage curves show a mode of 4, with a reduced proportion of southerly species compared with the present flora, but a maintained, perhaps slightly increased, proportion of northerly species. But there is still a majority of cold stage taxa in the classes widespread today, as with the E/W analysis.

The E/W indices (Figure 11.2)

The patterns of the curves for the present and cold stage floras are similar, except for the reduction of westerly species in the cold stage flora. The bulk of the flora in both floras (just over 50 per cent) is in the same modal classes (4–5), indicating the survival in the cold stages of present-day tolerant and

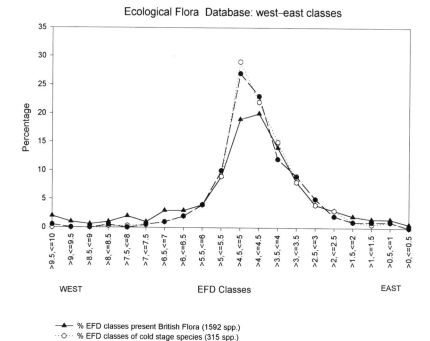

Figure 11.2. Analysis of the west–east distribution of the distribution classes of the Ecological Flora Database.

widely distributed species, but depletion of a westerly element. A converse increase in continental species is not so apparent.

This comparison of the present and cold stage floras, though the taxa totals used are very different and the cold stage flora is incompletely known, demonstrates the survival in cold stages of plants of wide distribution in Britain today, with indication of reduced westerly and southerly elements in the cold stages.

Other groupings of distribution in the PLANTDS1 table

Flora Europaea geographical groups

The *Flora Europaea* gives a summary in a short phrase of the distribution of each species within Europe. The PLANTDS1 table (FE field) attempts to list these for the cold stage species, giving an abbreviation (table 4.1) for these summary descriptions of geographical distribution within Europe. Table 11.3 shows the number of species ('a' and 'da') in each category where this is single.

Analysis of the *Flora Europaea* distributions relates the cold stage flora of the British Isles to the wider area of Europe. Allocation to the categories in Table 11.3 is not always straightforward and some simplifications have been made. The results give a view of the cold stage flora in a European context. The largest class of species (A, 221 spp.) is widespread in Europe at the present time. Many are widespread except at the geographical extremes. A smaller class (B, 91 spp.) is confined to areas of Europe, particularly northern and central areas. Species of arctic, subarctic and montane areas (C, 27 spp.) are fewer, as are principally montane species (D, 23 spp.).

The enumeration in Table 11.3 is striking in showing the variety of present distributions of the cold stage species represented fossil. As with the analyses above, most of the species are presently widespread, in this case in Europe. But the large number of species which have a narrower distribution in various parts of Europe, from the Arctic southwards, show the relation of the cold stage flora to the present flora, and how members of the cold stage flora are now in regions of varied climate and environment, an evolution of ranges following climatic change and the spread of forest or other vegetation formations under temperate stage conditions.

Widespread plants

This field lists the cold stage species described as widespread by Clapham, Tutin & Moore (1989), with support from the dot-maps of the B.S.B.I. *Atlas of the British Flora* (Perring & Walters 1982). These species occur rather widely east to west and north to south. The W category denotes widespread and more abundant species, the X category widespread but less abundant, more local, species and the Y category widespread coastal species. The numbers of species in each category are: W, 66; X, 78; Y, 15. The total is 159 out of 383 species ('a' and 'da' in the use field of the database), confirming the high proportion of present-day British widespread plants in the cold stage flora, as also indicated by the EFD analysis. Preston & Hill's (1997) floristic analysis annotates species which are widely naturalised outside the native range (the wid field in the PLANTDS2 Table). 115 out of 367 'a' and 'da' taxa are in this category, indicating a high proportion of recorded cold stage species are nowadays widely naturalised, a reflection of the capability of these plants to spread.

Table 11.3. *Numbers of species in* Flora Europaea *geographical groups*

Geographical groups	No. of species
A Widespread in Europe	
A1 Throughout most or almost all of Europe	79
A2 .... but not in extreme north or south	17
A3 .... but rarer or not in extreme north	35
A4 .... but rarer in north	10
A5 .... but rarer or not in extreme south	12
A6 .... but rarer in south	22
A7 .... but rarer in Mediterranean	23
A8 .... but only in mountains in the south	9
A9 .... but rarer or not in the east	1
A10 .... but rarer or not in the north-east	2
A11 .... but rarer or not in the south and east	1
A12 .... but rarer or not in the south and north-east	1
A13 .... but rarer or not in south-west and south	2
A14 .... but rarer or not in the south-east	5
A15 .... but rarer or not in south-west	2
Total	221
B Areas of Europe	
B1 North	10
B2 Fennoscandia	1
B3 West	6
B4 Central	3
B5 North, central	12
B6 North, east, central	2
B7 North, west, central	5
B8 North, north-east, central	1
B9 North-east	4
B10 North, west	5
B11 North-west, central	2
B12 North-west, west	1
B13 Central, east	2
B14 Central, west	10
B15 South, west	1
B16 South, west, central	18
B17 South, central	3
B18 South, central, east	1
B19 South, south-central, west	1
B20 South, south-east	2
B21 South, south-east, west	1
Total	91
C Arctic, subarctic, montane	
C1 Arctic Europe	4
C2 Arctic, subarctic	6

Table 11.3 (*cont.*)

Geographical groups	No. of species
C3 Arctic and mountains to south	2
C4 Arctic, subarctic and mountains to south	8
C5 Mountains of Norway, Sweden, Iceland	1
C6 Mountains of Fennoscandia, Scotland	2
C7 Mountains of central Europe	2
C8 Mountains of central Europe, Pyrenees, Scotland	1
C9 Mountains of south-east Europe	1
Total	27
D Montane and wider	
D1 North and mountains to south	15
D2 North, central and mountains to south	6
D3 North-west and mountains to south	1
D4 Mountains of central Europe, to north and south	1
Total	23

Arctic, Subarctic

The definition of these categories is always difficult (see Chapter 1). Northerly ranges may change with longitude (e.g. continentality), and only a small percentage of vascular plants in arctic floras are confined to that area (Major 1980). The fields list species whose present distribution reaches to the arctic (A) and/or subarctic (S); the data are is mainly from Clapham, Tutin & Moore (1989). Of the 383 species in the PLANTDS1 table, 185 (53 with single records) extend to the arctic and subarctic and 47 (13 with single records) to the subarctic. None is exclusive to the arctic. The remaining species do not have such a northerly distribution in Europe. The arctic field also lists alpine species (AL), of which there are three. The analysis shows the large number of cold stage species which presently extend into the arctic and subarctic, again demonstrating the present wide distribution of many cold stage species.

Non-British species

A listing of non-British species can only be a minimal list, because of the problems of identification of fossil taxa apparently not British, as discussed in Chapter 6. There may well be a bias towards the identification of species of British genera, with unidentifiable taxa, including taxa of identified genera, possibly non-British, excluded from the species lists.

A total of 23 species, all recorded as macroscopic remains, is listed in the non-British field of the PLANTDS1 table. Certain of these are associated with Devensian interstadials, for example, *Bruckenthalia spiculifolia* and *Picea abies*. The majority are terrestrial herbs, but the most abundantly recorded are the damp ground/aquatic species *Ranunculus hyperboreus* (in 28 samples) and *Potamogeton vaginatus* (13 samples). Eleven of the other taxa are recorded only once, the remainder between two and six times. Sparse as the evidence is, it does not appear that non-British species played an important part in the flora.

Distribution in time

The distribution of taxa in time is shown in the database in two ways. The MSAMSTAG and PSAMSTAG tables list the number of samples and sites containing respectively macroscopic remains and pollen/spores of each taxon in each cold stage/substage, so giving a survey of the presence of each taxon in time. The TAXATIME table summarises the occurrence of taxa, as macros or pollen/spores in cold stage stadials and interstadials, as well as in temperate stages and the present flora.

Records of cold stage taxa

The record is far better in the Devensian than in the previous cold stages, with far greater numbers of samples and sites than the earlier cold stages (Table 5.3). Knowledge of the flora of the Devensian is correspondingly much greater than in the earlier cold stages, as seen in Table 8.3 giving the number of taxa in stages/substages. Variation in the taphonomy of assemblages and the likely variation in the floras themselves, in terms of the patchiness of the vegetation or changes caused by climate or other environmental factors, make comparisons of floras at different times very difficult, apart from the distinction between stadial and interstadial floras.

In the question of the survival of the flora of the British Isles during the cold stages, the evidence from the Devensian radiocarbon-dated sites (Table 5.5) indicates survival of a rich flora in the periglacial area through the cold stage, including the time of maximum ice extension in the Late Devensian.

The richest macroscopic floras are those associated with fluviatile environments. Most are Devensian, with few from earlier cold stages. Of these latter, Broome (Wolstonian) and Ardleigh (Cromer Complex cold stage) are outstanding. If these floras are compared with Devensian floras of similar taphonomy, it is seen that they are very similar, with many taxa common to them (e.g. *Armeria, Betula nana, Carex,* Caryophyllaceae (*Cerastium*), many Crucifer species (*Draba, Erophila*), *Glaux,* Gramineae, *Hippuris, Linum perenne, Potentilla anserina, Salix*). Thus a typical cold stage flora extends back at least to the Cromer Complex cold stage. The records from yet older sites, such as the Beestonian sites (BV, BAM, BA-B), with a different taphonomy, show a poorer macroscopic flora but one with similar taxa (e.g. *Armeria, Betula nana, Carex,* Gramineae, *Potentilla anserina, Thalictrum alpinum, Saxifraga oppositifolia*), together with high percentages of Gramineae and Cyperaceae pollen.

Such cold stage assemblages as these evidently have appeared at our latitude in several cold stages, back to the Early Pleistocene. There is insufficient stratigraphic or floristic data to say whether any particular assemblage characterises a particular stage, as may be possible with temperate stage forest histories. Some differences are seen, e.g. *Helianthemum* pollen is not recorded before the Late Anglian, *Helianthemum canum* macros not before the Wolstonian, *Nymphaea* pollen not before the Devensian, Leguminosae pollen not before the Late Wolstonian, *Caltha* achenes not before the Early Wolstonian, *Onobrychis viciifolia* macros not before the Devensian. But many more data must be acquired to substantiate such differences.

Temperate taxa

The temperate field in the PLANTDS1 table lists the 'a' and 'da' species which have been found fossil in temperate stages of the Quaternary. The categories are: I, recorded in pre-Flandrian temperate (interglacial stages); IC, recorded in the Cromer Forest Bed Series (possibly temperate); F, recorded in the Flandrian temperate stage; A, recorded in an archaeological context in the Flandrian. The records used are those given by Godwin (1975).

Of the total of 383 species in the PLANTDS1 table, 269 (and 13 doubtful identifications) have been recorded in temperate stage sediments in the Quaternary. The totals in each category above are listed in Table 11.4. This is perhaps an unexpectedly high proportion of identified cold stage species. The number reflects the wide-ranging nature of many cold stage species, discussed above. For example, Godwin (1975) noted that several species of *Ranunculus* had been recorded in cold and temperate stages, including the Flandrian clearance periods, and considered their persistence through forested periods as a result of their association 'with a strong preference for river- or stream-side situations and damp habitats generally where forest dominance would be incomplete'.

Table 11.4. *Records of cold stage species in temperate stages*

Categories: I   pre-Flandrian temperate stage
              IC Cromer Forest Bed Series (possibly temperate)
              F   Flandrian temperate stage
              A   archaeological context in Flandrian

Category in PLANTDS1 table	No. of species	No. of species recorded once
I	23	8
IC	3	1
I, F	108	16
I, F, A	58	10
I, A	21	9
IC, F, A	1	1
F	21	5
F,A	11	5
A	14	11
IC, I?, F, A	1	–
I, F?	1	–
I?, F	2	2
I?, F, A	1	1
I, F, A?	2	–
I?, A	2	1

Conclusions

The study of present distributions of members of the British flora within Britain and within the wider area of Europe has usually had an objective of discerning the origin of the flora in geological time. Many geographical groupings have been distinguished. Species have been grouped into geographical entities and the origins of the group sought in past climatic change, as with the discussion by Forbes (1846). Such views parallel the idea that plant communities can be considered as entities which have a discrete history, a Clementsian view. The alternative, individualistic, Gleasonian view, is that a plant community should be considered as an entity bringing together species each having their own particular history.

The individualistic view applied to geographical groupings means that each species of a group will have its own particular history. Matthews' division of the Arctic-Alpine element into 'Historical Northern' and 'Historical Tertiary' species moves towards this view. With the increase of knowledge of the present distribution of species (e.g. the Ecological Flora Database), of their ecological physiology, of their variability and genetics, and of their fossil record, the individualistic view must come to the fore, naturally following earlier views of the origin of groups.

An important feature of the analyses described above is the large number of widespread species in the cold stage flora, associated with the large number of variable species (Chapter 9). Cold stage conditions evidently give the opportunity for many plants, especially short-lived species, to thrive in the open environments. This is in great contrast to conditions in the temperate stages, where long-lived forest trees become established and dominant in our latitude. Nevertheless, many cold stage species thrive in temperate stages, as shown above, in disturbed or marginal habitats, a tribute to their ability to respond to changing environments.

The analyses above have used the total of cold stage taxa. The taxa in each stage or substage could be analysed in the same way, to reveal variations between or within cold stages. But the poor record of pre-Devensian floras at present makes this a difficulty. At least the analyses above point out the possibilities of analysis, as well as providing a basis for discussing the interpretation of past climates.

12
The vegetation: types and their flora

Since the study of plant remains in Quaternary sediments started in the nineteenth century, relating fossil records of plants to the originating vegetation has naturally been a major aim. At first, there was the apparently simple relation of finds of macroscopic remains to major vegetation types, for example, the finds of plants now with an arctic distribution used to characterise glacial floras. With the advent of pollen analysis early in the twentieth century, applied especially to forest history of the last 10,000 years (Flandrian, Holocene), periods of dominance of particular assemblages of forest trees came to be identified, with the sequence now so widely recognised in Europe (e.g. Berglund *et al.* 1996). With increasing knowledge of pollen morphology, these investigations were pursued back to the more tree-less times beyond 10,000 years, to the late-glacial period of the last cold stage and to the times we are considering here.

Much has been learned of the relation between pollen deposition and forest composition, in terms of the pollen production and dispersal characteristics of tree genera. These matters are fully discussed in standard texts on Quaternary plant palaeoecology, e.g. Birks & Birks (1980), Berglund (1986). Pollen analysis has taken precedence in these studies because of the abundance and accessibility of the pollen evidence and the possibilities of quantitative analysis.

In the times before forests became dominant and replaced the mainly herbaceous vegetation of the cold stage stadials, the recognition of vegetation types via pollen analysis is far more difficult, even though it may be aided by studies of macroscopic remains. Such studies of macroscopic remains have not been seen to be so relevant to the periods of forest dominance, though they become again very significant in the times of forest clearance associated with human settlement.

There are two major difficulties in the interpretation of the cold stage record in terms of plant communities. The first concerns the use of present-day analogues. Analyses of recent pollen rain in relation to types of arctic vegetation have shown that spectra from the Arctic (with e.g. *Oxyria*) may be distinctly different from late-glacial spectra (with much *Artemisia*) in more southerly locations in northern Canada (Ritchie *et al.* 1987), illustrating the difficulty of using modern arctic spectra for the interpretation of cold stage spectra. This is hardly surprising in view of differences of latitude and climate and the consequences for the constitution and structure of the vegetation and the biology of the species concerned.

The second difficulty concerns the variety of taphonomy of the assemblages, of the habitats and of the mosaics of cold stage stadial vegetation which contribute to the record of pollen and macroscopic remains. Well-defined plant communities in the present Arctic can be widespread or form intricate mosaics (e.g. Batzli 1980; Sheard & Geale 1983). The pollen record amalgamates regional and local pollen production, while the macro record indicates more local occurrence of species. Both should be taken into account in attempting to define the originating communities. The pollen evidence can give rise to the identification of broad categories of vegetation, though more detailed or quantitative interpretations are difficult (see e.g. McGlone & Moar 1997). Since there are many more specific identifications based on macroscopic remains than on pollen identifications, local though broad communities can be recognised from these, as discussed in Chapter 10, though again quantitative interpretation is not feasible. The best case is where aquatic species in an environment defined by sediment can be clearly grouped as a community. But terrestrial species assemblages are more affected by processes of taphonomy and are thereby confused.

The approach taken here can be thought rather similar to that of Tansley (1911) in his classic account of *Types of British vegetation*, distinguishing types of vegetation and their component taxa.

Before venturing an account of the vegetation in more detail, it is worth considering some biological points which arise in comparing present northern vegetation and flora with cold stage vegetation at lower latitudes. The comparison draws attention to the character of differences which play an important part in plant life, liable to affect notably productivity and reproductive capacity.

A key consideration concerns seasonal differences, involving concomitant temperature and precipitation regimes. These are imposed on the difference in the length of the growing season, much shorter in the Arctic (say 6–9 weeks, but very variable) compared with the potentially much longer season of several months at our latitude. The soils of the Arctic, often wet in the thaw period or through poor drainage on permafrost, are at low temperatures, limiting growth. Warmer spring temperatures at lower latitudes, and drier soils, will encourage, for example, shoot growth of herbs, including grasses and sedges, while warmer air temperatures will increase promote photosynthesis and productivity. Chapin & Shaver (1985) give a comparison of annual productivity of graminoids in the arctic and temperate grassland, showing that although relative growth rates may be similar, productivity of the temperate grassland is much greater, explained by longer growing season and greater quantity of over-wintering green tissue. Haviland (1926) notes that the warmth of the spring favours growth of grass in the steppe. As with *Onobrychis viciifolia*, increased warmth may also increase nectar production, to the benefit of entomophilous species.

A further major factor affecting distribution of arctic plant commmunities is soil moisture. Herbaceous cold stage communities will be affected in the same way, but with seasonal variations imposed, a greater complexity of conditions and vegetation might be expected.

These differences will also produce changes in the structure of vegetation, which may affect pollen or seed distribution. Forbs in the Arctic often flower at low levels, where it may be warmer; later, scapes elongate, so enhancing seed dispersal. In lower latitude grassland, forbs have greater stature, improving the possibility of pollen dispersal.

Pollen deposition rates in the Arctic compared with grasslands further south (Ritchie & Lichti-Fedorovich 1967; Ritchie & Cwynar 1982) show a great difference in pollen productivity with latitude, which have to be taken into account in interpretation of cold stage pollen spectra. Closed herbaceous communities give rise to higher pollen production, but as with forest pollen production, structure must be important in determining the representation of the component species.

The following account of cold stage vegetation is divided into two sections dealing with interstadial and stadial vegetation. The former includes the Chelford and Brimpton Interstadials and the Upton Warren Interstadial Complex of the Devensian, where interpretation of the first two is facilitated by the knowledge of concepts about forest history. The latter includes the remainder of the cold stage record, the bulk of which is Devensian in age.

Interstadial vegetation

A definition of the term 'interstadial' was discussed in Chapter 1, and the palaeobotanical usage suggested by Jessen & Milthers (1928) (Figure 1.1) was adopted for palaeobotanical studies such as this analysis of the cold stage flora. However, the term has also been used to describe intervals thought to be less severe than stadial on other biological evidence, particularly, in north-west Europe, fossil insects. In Britain a number of sites have yielded evidence from flora and fauna of interstadial conditions in the Devensian cold stage, preceding the Devensian late-glacial (see SITE CHAR table). They are as follows (in order of increasing age; site numbers in parentheses):

> Upton Warren Interstadial Complex: Earith (41), Four Ashes (31), Isleworth (10), Ismaili Centre (14), Kew Bridge (13), Upton Warren (30), Twickenham (9).
> Brimpton Interstadial: Brimpton (3).
> Chelford Interstadial: Beetley (62), Brimpton (3), Chelford (65), Four Ashes (31), Wretton (43).

Evidence of flora and fauna supports the designation of the Chelford and Brimpton sites as interstadial. On the other hand, evidence from insects distinguishes the Upton Warren Interstadial Complex sites, with the flora showing no great distinction from stadial floras. A second woodland interstadial at Wretton, predating the Chelford Interstadial at that site, was proposed by West *et al.* (1974), but the samples have been excluded from the data base because of the abundance of reworked plant remains. The vegetation of these interstadials can now be considered.

Chelford Interstadial

The first Devensian interstadial site described in detail was Chelford (Simpson & West 1958), which became a 'type' site. There was little vegetational change through the thin organic detritus muds at this site, the plant macroscopic remains and pollen indicating the presence of forest with *Betula, Pinus* and *Picea*, a ground flora with *Calluna* and *Empetrum*, and wide-ranging and/or northern swamp plants. The *Pinus* pollen frequency distribution is shown in Figure 8.4. Simpson & West (*loc. cit.*) drew a comparison with forest of the Fennoscandian conifer-birch region (say 63°–69° N). The presence of *Lonicera xylosteum* pollen, a characteristic shrub of this forest, was notable.

At Beetley (Phillips 1976), a similar pollen assemblage was recorded in sediments of a shallow pool and correlated with the Chelford site. There were particularly abundant needles of *Picea abies* cf. ssp. *obovata*. The record of macroscopic remains is far more extensive, with elements of heath and acid wetland communities, including *Bruckenthalia spiculifolia*, *Betula nana*, *Erica* sp., *Eriophorum vaginatum* and *Sphagnum*. It was succeeded by a pollen zone which showed a reduction in *Pinus* pollen percentages, with an increase in pollen of ericaceous shrubs and herbs, especially Gramineae pollen. A transition to subsequent stadial conditions, with high frequencies of non-tree pollen, is clearly shown in the pollen diagram.

At Brimpton (Bryant *et al.* 1983) fine sediments within fluviatile gravels yielded pollen spectra again similar to those at Chelford, with heath and acid wetland shrubs and herbs. The macro flora included abundant needles of *Picea abies* cf. ssp. *obovata*. At Four Ashes (Andrew & West 1977), a single level of organic sediment in fluviatile gravels gave a pollen spectrum similar to those from Chelford.

At Wretton (West *et al.* 1974), organic sediments of small shallow pools within a fluviatile sequence contained a tree pollen assemblage similar to that at Chelford. At one of these sites (Wretton WG) two pollen assemblage zones were distinguished, a *Pinus-Betula-Picea* biozone followed by a *Pinus-Betula-Picea-Calluna-Sphagnum* biozone, indicating expansion of wet heath communities in the area.

The Chelford assemblages show the extension of birch-pine-spruce forests into Britain during the Early Devensian. Though the *Picea* pollen pecentages are low, leaves of *Picea* are abundant, with in identification to ssp. *obovata* and cf. ssp. *obovata* at some sites. Understorey shrubs such as *Bruckenthalia*, *Calluna*, *Vaccinium myrtillus* and *Empetrum* are also present. Unlike the more detailed long Early Weichselian sequences in continental north-west Europe, the Chelford Interstadial sequences are much more fragmentary. Only the sites at Beetley and Wretton show vegetation change, both with expansion of ericaceous shrubs and herbaceous taxa following the major forest development.

The Chelford assemblages also include taxa more characteristic of open ground conditions, with wide edaphic preferences, and of shallow water bodies. All the sites are located in or near fluviatile sediment sequences, and habitats satisfying these requirements are likely to have been on the floodplains, in contrast to regional forests. Such environments would have provided havens for stadial taxa within regional forest.

Looking at the outline development of vegetation from the end of the preceding temperate stage (Ipswichian), the Early Devensian stadial vegetation is succeeded by the spread of the Chelford forests, later associated with shrub communities with *Calluna* in particular. This parallels developments in continental north-west Europe, where a similar forest replaced stadial vegetation, with heath communities later expanding, in the Early

Weichselian Brørup Interstadial (see Behre 1989 for a survey of vegetation at this time in north-west Europe).

Very few taxa have been recorded in the Chelford Interstadial but not in the stadial assemblages. They include pollen records of *Gentiana pneumonanthe* and *Lonicera xylosteum*, and macro records of *Picea abies, P. abies* ssp. *obovata, Betula pubescens, Rubus fruticosus* and *Vaccinium myrtillus.*

Brimpton Interstadial

This interstadial is based on a single site at Brimpton (Bryant *et al.* 1983), where several samples of silt filling channels or at bar-tops in a fluviatile sequence yielded pollen spectra with more tree pollen than in underlying cold stadial sediments. Pollen of *Betula* and *Pinus* totalled 30–40 per cent, with only few grains of *Picea*, so differentiating the assemblage from the older Chelford Interstadial assemblage. Gramineae pollen was the major contributor to the non-tree pollen total, with taxa also indicative of grassland, heath, fen and aquatic habitats. It was suggested that the assemblage was derived from upland forest and more local habitats including grassland. Also present was a mollusc assemblage which contained thermophilous species indicating interstadial warmth, including species characteristic of damp grassland.

Correlation of the Chelford and Brimpton Interstadials

The Early Weichselian of north-west Europe is thought presently to include the two interstadials, the Brørup and Odderade Interstadials (Behre 1989; Behre & van der Plicht 1992; Donner 1995). The Chelford Interstadial has been correlated with the former, both showing forest assemblages with *Picea*. The Brimpton Interstadial has been correlated with the Odderade Interstadial. The tree pollen percentages are rather lower at Brimpton than in the Odderade assemblages; this may be an expression of climatic or migration gradients.

Upton Warren Interstadial Complex

This interstadial was first described at the Worcestershire site of Upton Warren (Coope *et al.* 1961; Coope *et al.* 1971), where insect faunas in Devensian fluviatile sediments indicated warm July temperatures, warmer than those indicated at other Devensian sites. Similar insect faunas have

Plate 1 A. Fluviatile Devensian gravels within a terrace, with a series of channels outlined by large-scale cross-stratification shown by thin black organic beds within fine bar-tail sediments. Section SAD at Somersham (West *et al.* 1999). Sp, spoil; P, diamicton; F2, Middle/Late Devensian gravels. **B.** Detail of bar-tail pool sediments in section SAD at Somersham. Coarse drift mud accumulation at base containing a rich macro flora and very poor pollen flora, fine inorganic laminated sediments above. Pencil 8 cm long (West *et al.* 1999). **C.** Late Devensian laminated pool sediments at section SY at Somersham. Organic drift mud (radiocarbon age 18.31 ka) with a rich macro flora but poor pollen flora concentrated in one horizon (K) within clayey sediments, overlying sand (J) (West *et al.* 1999).

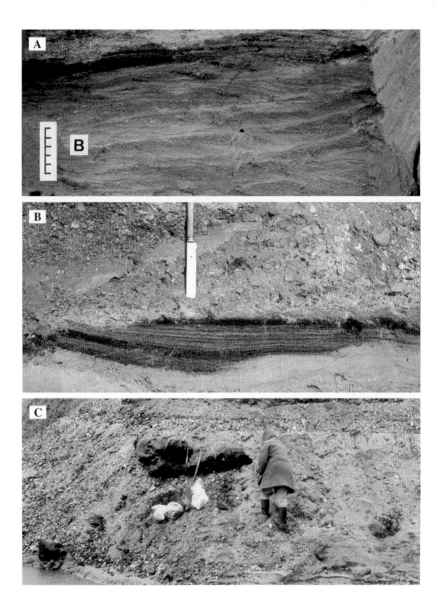

Plate 2. **A.** Wolstonian overbank sediments of section SAP at Somersham. Coarsening-up units (B) aggrading to the right are overlain by more organic laminated sediments with a rich macro flora and very poor pollen floras. Cm scale (West *et al.* 1999). **B.** Coarse laminated sand unit within Devensian fluviatile sands and gravels. The dark organic beds contain a very poor flora, probably a result of sorting in a more active flow regime. Section 16, Block Fen (West *et al.* 1995) Photo P.L. Gibbard. **C.** A massive thick organic bed filling a pool, probably an abandoned meander pool, in Middle Devensian gravels at Beetley (section RR, West 1991). A well-preserved pollen flora is present, but macroscopic remains are restricted in number, contrasting with the representation of pollen and macros in the bar-tail sediments.

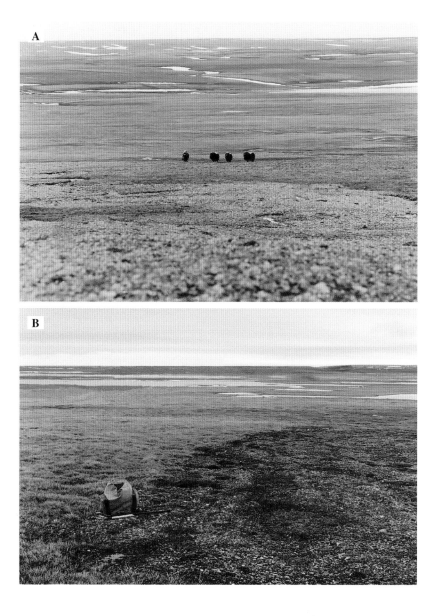

Plate 3   Examples of vegetation variation in the region of Polar Bear Pass, Bathurst Island, Nunarut, Canada.   **A.** View looking south over the mesic sedge communities widespread on the poorly-drained and heavier valley soils, supporting herbivores (musk oxen, centre), contrasting with polar desert of the drier rocky slopes of the foreground.   **B.** A sharp margin between mesic sedge communities of the valley and better-drained soils of the slopes with open plant communities.

Plate 4. Examples of vegetation variation in the area of Sachs Harbour, south-west Banks Island, N.W.T., Canada. Annual rainfall is relatively low and the floodplains are largely sandy. **A.** Sandy floodplain of the Kellett River, supporting a rich herb flora, with few erect shrubs. The pools contain sediment with a varying content of plant remains. **B.** On the southern side of the Sachs River, shrubs are associated with sand dunes and border mesic sedge communities in the valley, with drier conditions on the left hand slope. Here, the darker shrubs are *Arctostaphylos alpina*.

A

Plate 5. Examples of vegetation variation. **A.** Sachs Harbour, Banks Island. A sharp boundary between, on the right, closed and more stable plant communities with low *Salix* and *Carex* in a shallow valley and open communities on hummocky ground subject to frost-heaving and movement. **B.** The western bank of the Colville River, Alaska coastal plain, in the region of Umiat. Dunes with shrubs bordering the river channel, with closed herb communities and further open sandy ground intervening between the wet tundra with polygons and the river. An example of the great variation of vegetation in a small area on and adjacent to a floodplain.

Plate 6. **A.** Variation of vegetation in a valley on the southern edge of the Alaska coastal plain. Stream margins, better drained, show shrub communities, which also occur on some of the slopes. Wet and dry polygons are seen in the lowland area.
B. Herschel Island, Yukon, Canada. A side view of the mesic herb-rich vegetation at the time of grass flowering, showing the extension of flowering culms above the general level of herbs, aiding anemophilous pollen dispersal.

Plate 7. **A.** Plants of an alkaline basin, near Takhini Crossing, Yukon, Canada with *Suaeda maritima* and *Salicornia rubra*. *Glaux maritima* and *Linum perenne* are also present in this inland basin in the rain shadow of the St Elias range. Cm scale.
B. Sand dune plants adjacent to sand floodplain of Sachs River, Banks Island, N.W.T., Canada. *Cerastium beeringianum, Melandrium affine, Armeria maritima* and *Artemisia richardsoniana*.

Plate 8. **A.** *Saxifraga oppositifolia,* a cushion plant well able to survive in polar desert, here on a gravel plain, Bathurst Island, Nunavut, Canada. **B.** Vegetation of a marginal gravel plain of the Porcupine River, Yukon, Canada, with *Salix arbusculoides* and *Allium schoenoprasum*; also present *Polygonum alaskanum, Potentilla anserina, Sanguisorba officinalis, Hedysarum mackenzii, Galium boreale*.

since been described from Devensian floodplain sediments at many other sites (e.g. Coope & Angus 1975).

Plant remains have been analysed from many of these sites (listed above). The macro and pollen assemblages show the presence of herbaceous vegetation. There are high percentages of non-tree pollen, including in particular Gramineae and Cyperaceae pollen, as seen in the pollen frequency distributions for these taxa in Figures 8.2 (Gramineae) and 8.3 (Cyperaceae).

Unlike the evidence of distinct insect faunas, there is little substantial evidence of a distinct flora or vegetation. In the Thames sites with these faunas, there is some suggestion that a richer aquatic and marsh flora, together with tall herb communities (with *Anthriscus, Heracleum,* and Umbelliferae pollen), was significantly present, in contrast to an earlier time at the site (Gibbard 1985; Coope *et al.* 1997). However, the taxa concerned also occur at other times in the Devensian. It is to be noted in respect of climatic interpretation that such tall herb communities rich in Umbelliferae species are a feature of meadows of the continental areas of southern Siberia. Further analysis is needed, especially because of the very interesting contrast between the climatic evidence offered by the insects and plants, which is considered in Chapter 13.

Since no great differences have been established between the Upton Warren Interstadial floras and the stadial floras, both showing high frequencies of non-tree pollen, the interstadial assemblages are included in the discussion of stadial vegetation.

Stadial vegetation

In contrast to the small number of sites with evidence of interstadial vegetation, there are a very large number of sites for the analysis of stadial vegetation. The site details are available in the database tables, and rather than a site-by-site approach, a more generalised study is made. This surveys regional and local vegetation, considers the possibility of identifying vegetation types, and lists plants associated with particular habitats. Certain sites informative about local vegetation are discussed, and also periods which may show particular types of vegetation. Certain taxa of interest in relation to the vegetation are also noted.

Regional and local distribution of vegetation

Variation of regional stadial herbaceous vegetation can obviously be great. Two extremes are possible. One will be represented by the vegetation of plateaus such as those of the boulder clay of East Anglia, where similarities of soil conditions and exposure are likely to have contributed to a uniformity of vegetation, though areas may have been subject to periglacial frost action leading to patterned ground and habitat variety (Figure 10.1; Watt *et al.* 1966).

At the other extreme will be areas of very varied relief, such as those present in the young morainic landscape of Herschel Island (Yukon Territory), with a complex mosaic of vegetation types dependent on soil, drainage, aspect and permafrost depth (Smith *et al.* 1989), or the very varied habitats associated with the gravelly river floodplains of the cold stages. The proportion of each such type of landscape in an area will affect the fossil record via the taphonomy of the assemblages, so interpretation of assemblages needs consideration of the landscape. This is illustrated later by reference to particular sites. Figure 12.1 and Plates 3, 4, 5 and 6A illustrate this variety in the present Arctic.

The difficulty of identifying vegetation types from the fossil record varies according to the detail required. In studies of forest history, regional formations can usually be identified from the pollen assemblages, but the identification of herbaceous vegetation types and plant communities is more difficult. Schweger (1982) discusses this problem in relation to eastern Beringia vegetation. Communities are defined by their species content and abundance. But species found fossil can occur in several communities and the abundance of a species giving rise to fossils is not determinable from the fossil record.

Using the known occurrence of taxa in particular vegetation types at present (vegetation field, PLANTHAB table), an approach can be made through these data to assemble plant lists of particular vegetation types, remembering that plant communities different from those now recognised may have been present in the past.

However, many species occur in more than one vegetation type (Table 10.1), and this blurs the distinctions, particularly with the two most abundant and common stadial pollen taxa, Gramineae and Cyperaceae. Species of both families characterise a great variety of vegetation and habitat, dry and wet and in closed or open communities. The significance of these pollen taxa is clearly a basic issue in interpreting the vegetation, which can be assisted by knowledge of the species concerned and of the sedimentary environment.

Analysis of the vegetation can best start with a discussion of so-called steppe-tundra vegetation and consideration of the significance of the records of Gramineae and Cyperaceae. These two taxa are usually present

Figure 12.1. View of the western end of Polar Bear Pass, Bathurst Island, Nunavut, Canada, looking south. Polar desert in the foreground, with scattered *Saxifraga oppositifolia* and *Salix arctica*. The valley is occupied by largely closed communities with mesic sedge meadows and shallow pools, in great contrast to the drier slopes of the valley sides. Permafrost impedes the drainage in the valley and elsewhere, as do the more clayey soils (of marine origin) of the valley. Vegetation/soil stripes are visible on the lower slopes, intermediate between the polar desert and the more mesic lowland.

in substantial percentages in cold stage pollen diagrams, in varying proportion (Figure 12.2). In the discussion of vegetation which then follows, species are listed for each vegetation type in which they occur at present, so that species with wide preferences are noted under more than one heading. To provide more certainty in the interpretations, the lists usually only mention species which have been recorded in five or more macroscopic and pollen samples, so excluding rarely recorded species (the PLANTHAB table gives the complete lists); also excluded are taxa thought to be reworked or of doubtful provenance and non-British taxa. These lists are used for analysis in terms of the distribution categories of Preston & Hill (1997) (Table 12.1).

For each species the number of macroscopic (m) and pollen (p) records is given in parentheses. The sample numbers exclude records from the Brimpton and Chelford Interstadials of the Devensian, but include samples from the Upton Warren Interstadial Complex. The more detailed distribution in time of the taxa involved is given in the TAXATIME table, so that the

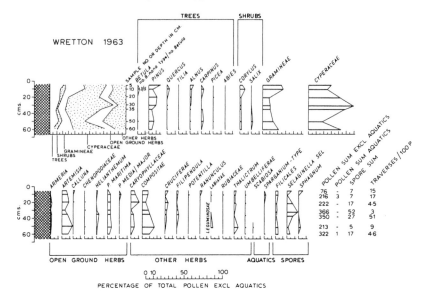

Figure 12.2. Pollen diagram from organic sediments of an Early Devensian meander cut-off pool at Wretton, Norfolk (site WH). The low frequency of tree pollen, the marked frequencies of Gramineae and Cyperaceae pollen and the variety of herb pollen types are characteristic of stadial pollen spectra from periglacial southern Britain (Sparks & West 1972).

occurrence of particular species in particular cold stages or substages could be further analysed.

Steppe-tundra, Gramineae, Cyperaceae

Steppe-tundra

Hibbert (1982) has described the history of the concept of steppe-tundra vegetation. Definitions of these physiognomic terms are discussed in Chapter 1. The use of the term implies that the vegetation has certain floristic and physiognomic characters of both steppe and tundra. This gives rise to a problem of interpreting contemporary climate or climates, discussed by Chernov (1985) in relation to both flora and fauna.

The occurrence in cold stage sediments of remains of vertebrates nowadays associated with the Eurasian steppe formation was described and discussed by Nehring (1890). The association gave rise to the proposition that during the last cold stage (and earlier cold stages) steppe-like conditions, with loess deposition, favoured the expansion of steppe species in Europe.

Table 12.1. *Number of species of the vegetation or habitat categories recorded in five or more stadial macroscopic and pollen samples in the floristic elements of Preston & Hill (1997)*

	Category in PLANTHAB table[a]											
	G	Me	Fl	D	B	H	Mo	W	Ma	Fe	A	MM
Major biome category												
1. Arctic-montane	1	–	3	–	–	–	1	–	–	–	–	–
2. Boreo-arctic Montane	4	–	4	1	1	1	3	–	–	–	–	–
3. Wide-boreal	2	1	1	–	–	–	–	1	2	1	–	–
4. Boreal-montane	2	–	2	–	–	–	–	–	–	–	3	–
5. Boreo-temperate	17	6	1	4	5	8	5	6	9	11	11	8
6. Wide-temperate	2	1	2	–	–	1	–	–	2	–	5	9
7. Temperate	8	2	–	1	1	2	1	1	8	5	8	5
8. Southern-temperate	3	–	–	–	2	–	–	1	3	3	4	2
9. Mediterranean	1	–	–	–	–	–	–	–	–	–	–	–
Eastern limit category												
1. Oceanic	–	–	–	–	–	–	–	–	–	–	–	–
2. Suboceanic	2	–	–	–	1	–	–	–	1	1	1	1
3. European	7	3	5	2	4	5	2	2	4	4	4	5
4. Eurosiberian	12	2	2	1	1	4	1	4	3	7	3	5
5. Eurasian	5	4	2	1	–	–	–	1	7	3	5	5
6. Circumpolar	14	1	4	2	3	3	7	2	9	5	18	8

[a] G, grassland, downs; Me, meadows; Fl, fell, rocks, etc.; D, dunes; B, bogs; H, heath; Mo, moors; W, woodland, scrub; Ma, marshes; Fe, fens, flushes; A, aquatic; MM, man-made (includes Cg, cultivated ground; Ds, disturbed ground; Fi, fields; Ro, road/waysides; Wa, waste places).

With the later development of Quaternary palaeobotany, cold stage pollen assemblages and macroscopic remains showed the existence of taxa and vegetation formations which supported these ideas. Thus pollen assemblages showed high frequencies of Gramineae pollen, with *Artemisia* also present in notable frequency, and macro assemblages contained species of continental distribution type. The evidence suggested grassland and a continental climate.

Intensive studies of the last cold stage biota in the Beringia region (Hopkins *et al.* 1982), adjacent to the Bering Straits, have added much very interesting detail to this concept of what has become variously known as the arctic steppe biome, steppe-tundra (or the reverse), and mammoth steppe (Guthrie 1990). The last term is in reference to a diverse fauna of large gregarious vertebrates (e.g. horse, bison, mammoth, reindeer, Saiga antelope) which is found associated with many sites of the last cold stage, and which is dependent on the vegetation of the time. The small vertebrate and insect faunas have added weight to the steppe connection (Matthews 1982).

However simple this concept may appear, there has been much discussion on the details of the evidence. To support a supposedly substantial grazing population, productivity of vegetation has to be sufficient, and an important open question is the degree of productivity of the vegetation indicated by pollen analysis. Pollen influx data may assist the answer, as may more exact dating of times when the fauna flourished (see e.g. Ritchie & Cwynar 1982; Colinvaux 1986; Ukraintseva 1986; Guthrie 1990; Schweger 1990, for description of stomach contents of some large grazers of the time). Nevertheless, the grazing fauna and the steppe elements of plants and animals give persuasive evidence of a particular vegetation formation, physiognomically steppe-like and productive, evidence which has been found from Beringia westwards to western Europe.

This is not to say that unbroken extensive areas of steppe-tundra existed at times. A mosaic of vegetation, dependent on the usual edaphic, climatic (mega- and micro-) and other factors will obviously have been present (described e.g. by Schweger 1982; Young 1982; mapped for the Late Valdai Glaciation by Grichuk 1984), and one of these factors may have been grazing pressure, as suggested by Zimov *et al.* (1995). The non-zonal presence of vegetation with steppe species within taiga and tundra in Siberia and Alaska may possibly be a relict of cold stage steppe-tundra. Such areas include, for example, the south-facing slopes of pingos on the Arctic slope of Alaska (Walker *et al.* 1991) and Wrangel Island on the Chukchi shelf of Siberia (Yurtsev 1982; Walter 1985)). The flora of the latter includes *Linum perenne*, a common cold stage plant of Britain.

These comments provide a necessary background to the interpretation of cold stage vegetation in Britain. The over-riding frequency of Gramineae pollen, the presence of continental species (Chapter 11), and of the large grazers listed above, parallel observations in continental Eurasia and eastwards. Superimposed on this is the diversity of habitats and vegetation indicated by the flora.

Gramineae

Gramineae is the most common pollen taxon in the cold stage flora, with high frequencies in the stadial pollen assemblages (Figure 8.2). The same taxon has the third most frequent records of macroscopic remains, to which can be added a number of records of genera and species. The abundance of records reflects the anemophily of the family and the diversity of habitats the species occupy.

Compared with taxa of Cyperaceae, Gramineae genera and species are more difficult to identify from macroscopic remains. Disappointingly few have been identified in the cold stage flora, with only six generic identifications and three specific. All are of terrestrial genera, except for *Glyceria*, with both a pollen type and caryopses identified. The terrestrial taxa are *Festuca* (*F. rubra*, *F. halleri*), *Poa*, *Anthoxanthum* (*A. odoratum*), *Agrostis* and *Alopecurus*, with a number of cf. identifications of these and two other genera or species (*Elymus*, *Hordeum*). These genera occupy a wide range of habitats from grassland to disturbed and open ground, such as are well described on Herschel Island (Smith *et al.* 1989).

Gramineae pollen percentages in the stadial assemblages are high (Figure 8.2), often higher than the percentages reported from the Canadian mid and high Arctic by, amongst others, Ritchie *et al.* (1987). The problem in interpretation of these frequencies is the ecological width of the family and the question of over-representation. Over-representation is likely to occur, by virtue of both the flowering height of the grass culm and the low stature and entomophily of other herbs (Plate 6B). On the other hand, Iversen (in Degerbøl & Iversen 1945) presented data from the inland region of Godthaab Fjord in Greenland which indicated grass pollen was under-represented in recent lake muds compared with the percentage composition of Gramineae in the local vegetation; local pollen productivity in that area may not parallel production in lower latitudes.

Pollen concentration or influx numbers here would help interpretation, but in the absence of such data (acquisition hindered by the problem of measuring sediment accumulation rates), the high frequencies of Gramineae pollen, supported by the frequency of macroscopic remains, indicates the importance of the family in the regional and local vegetation. Regional grassland is suggested. Grassland as a formation occurs at a wide range of latitudes, as discussed in Chapter 1, from polar steppe to lower latitudes, which leads to further difficulties in the interpretation of the pollen assemblages in terms of vegetation.

Further, grassland is typified by a wide variety of Gramineae genera and species (Coupland 1979; French 1979), and with a poor fossil record of Gramineae taxa, identification of a particular type of grassland is impossible at present. The lack of knowledge here also hinders the interpretation of the relation between the grassland and the herbivore population, since grassland varies considerably in productivity. A knowledge of the presence of C_3 or C_4 species might also help, since the latter are more common in hotter and drier climates.

Cyperaceae

The taxon Cyperaceae has the most frequent cold stage records of macroscopic remains and the second most frequent (after Gramineae) of pollen records, with high frequencies in the assemblages (Figure 8.3). This must be partly due to the anemophily of the family and the abundance of species which occupy aquatic, semi-aquatic, fen and marsh habitats, where preservation of macroscopic remains is favoured. But species of the family occupy a wide range of habitats, from aquatic to fully terrestrial, and questions arise about the distribution of species in different habitats. Most of the species recorded in some frequency are associated with aquatic or wet habitats, but some less frequent or rarely-recorded taxa can occur in less wet habitats and may be associated with regional vegetation (e.g. *Carex panicea, C. flacca*).

Modern pollen spectra from lakes of tundra regions in Canada show substantial Cyperaceae pollen percentages (Ritchie *et al.* 1987), but the contribution to these from the many aquatic and terrestrial species of the region is unknown. Surface samples from the tundra of Banks Island, N.W.T., show high frequencies in wet floodplain areas with *Eriophorum* communities, less high in a drier upland area with lower vegetation cover (Schweger 1982). As in this case, high Cyperaceae frequencies in cold stage pollen assemblages are usually asssociated with wetland or aquatic sedimentary facies.

In relation to a concept of cold stage stadial grassland, it must be noted that present steppe grassland of eastern Europe and eastwards includes a number of species of *Carex*, including the British *C. humilis* as an important species. However, as with the modern Canadian Cyperaceae percentages, it is impossible to apportion the contributions of aquatic and terrestrial species. At one Devensian site, Beetley WW, 20–30 per cent Cyperaceae pollen occurs with frequent pollen of drier ground herbs near the base of a shallow pool filling, rising when frquencies of the latter fall later in the filling of the pool by organic sediment. At another Devensian site, Block Fen site 15, there are variable and substantial Cyperaceae pollen percentages again associated with frequent pollen of drier ground herbs. These percentages may be an indication of regional Cyperaceae in grassland. Conditions are not favourable to the preservation of fossils of terrestrial taxa of Cyperaceae, as seen in the fossil record, but it seems reasonable to suppose that a proportion of Cyperaceae pollen in the stadial pollen spectra originated from terrestrial regional vegetation.

Vegetation, taxa, habitats

Polar desert

It would seem doubtful if any plant community which could be described as indicating regional polar desert could develop at the lower latitudes we are considering. True polar desert is characterised by cushion plants, tussock grasses and lichens, with very low cover percentages. Soil development is limited and there is little soil moisture except in the spring. Carbonate encrustations, salt efflorescences and saline conditions are associated with dry summers and windy conditions.

Some of the conditions controlling plant life of present polar desert may be active in cold stage stage environments, such as dry summers and wind. These, aided by the reduction of plant cover by deposition of aeolian sediments as in dunes, or by the presence of rock fields, may well produce desert-like conditions in places, but the fossil record shows that floras were generally much richer at the lower latitudes, unlike the restricted floras of present polar desert. A similar enrichment of the flora of the high Arctic where conditions are more favourable is seen in the existence of the so-called high Arctic oases, such as the Lake Hazen valley in northern Ellesmere Island. The difference between pollen spectra from polar desert and from cold stage spectra at lower latitudes with a richer and more thermophilous flora was pointed out by Funder & Abrahamsen (1988).

Grassland and meadow

Pollen assemblages with a dominance of Gramineae pollen may originate from a variety of grassland communities, temperate, boreal and arctic. The distinction can be drawn, for example, between grassland of treeless areas associated with continental climates such as the steppe, and grassland associated with more mesic conditions of alluvial areas. The latter have been termed meadows, but this term is also used for anthropogenic grassland cut for hay or grazed, a reminder that conditions other than climate can control the presence of grassland. Fire, as well as grazing, is also a possible factor; both are possible in cold stages. The term, however, is useful in a descriptive sense, regardless of possible origins, since tall-grass or tall-herb communities are important vegetation types in temperate and boreal areas. There are all gradations between mesic meadow vegetation and steppe, along a gradient of increasing aridity, succinctly described by Walter (1985). Such a gradient may be regional or local. Printz (1921) described a marked local gradient of this type in his account of the vegetation of the area of the Abakan River in southern Siberia, with a contrast between the steppe and the rich meadows of the river valley.

A similar kind of gradient of late Wisconsin vegetation in eastern Beringia was visualised by Schweger (1982). He proposed

two broadly defined vegetation types, an upland xeric tundra dominated by sedges, *Artemisia* and grasses, as well as numerous but less abundant herbaceous meadows (or tundra) with frequent willow shrubs,

the latter more mesic and in the lowland valleys.

The two categories of grassland and meadow will be considered in relation to the fossil record.

Grassland

The abundance of macroscopic and pollen records of Gramineae has already been discussed. Against this background there are the records of taxa of grassland (or downs) labelled G in the <u>vegetation</u> field of the PLANTHAB table. These form by far the largest vegetation category, with over 100 taxa. Nearly half of these occur in other vegetation categories, which may help to explain the abundance of taxa. The number of taxa may also appear surprisingly large if they mostly belonged to the drier categories of grassland where potential preservation of fossils is low. But taxa of dry grassland, e.g. *Artemisia* (p425) and *Helianthemum canum*, may also occur in drier soils of floodplains, where preservation is more likely. Interpretation of grassland types is also hindered by the small number of grass species identified, in contrast to the large number of identified forb taxa associated with grassland.

A large number of species in the grassland category have been frequently found in stadial samples. These include *Selaginella selaginoides* (m69, p26), *Polygonum viviparum* (m31), *Rumex acetosella* (m24, p26), *R. acetosa* (m8, p26), *Cerastium arvense* (m29), *Thalictrum minus* (m11), *Sanguisorba officinalis* (m1, p22), *S. minor* (m4, p5), *Potentilla anserina* (m83), *Medicago sativa* ssp. *falcata* (m10), *Onobrychis viciifolia* (m9), *Linum perenne* and *L. perenne* ssp. *anglica* (m42, p4), *L. catharticum* (m8, p4), *Euphorbia cyparissias* (m10), *Helianthemum canum* (m45) (*Helianthemum* (m7, p97)), *Armeria maritima* (m72, p41), *Plantago maritima* (m5, p120), *P. lanceolata* (p18), *Valeriana officinalis* (m5, p58), *Scabiosa columbaria* (m6, p4), *Campanula rotundifolia* (m60), *Artemisia* (p425), *Centaurea nigra* (p20), and *Allium schoenoprasum* (m17).

Many of the species prefer dry calcareous soils, which may have been widespread on the Chalk or other limestones of south-east England, on sandy soils nourished by loess and on cover sands of floodplains or elsewhere. The difficulty remains of identifying any narrow vegetation units. The taxa list has to be read in association with the background of high Gramineae pollen frequencies.

Meadows

Twenty-two taxa are labelled meadow plants in the PLANTHAB table, but most show wider preferences. The list includes *Ranunculus repens* (m28), *R. acris* (m14, p2), *Thalictrum flavum* (m38), *Filipendula ulmaria* (m17, p5), *Pedicularis palustris* (m9), *Valeriana dioica* (m3 p2), *Succisa pratensis* (p13), *Achillea millefolium* (m12), *Leontodon autumnalis* (m13) and *Taraxacum officinale* (m13). To these should be added *Polemonium* with few pollen records.

Of the species with five or more sample records in the grassland amd meadow categories, the majority are in the Boreo-temperate and Temperate categories of Preston & Hill (1997), with a smaller number in categories 1 to 4 and 6 (Arctic-montane, Boreo-arctic Montane, Wide-boreal, Boreal-montane and Wide-temperate) and 8–9 (Southern-temperate and Mediterranean), emphasising that the flora is mostly temperate in distribution at present (Table 12.1).

Fells, rocks, screes, ledges, crevices, barrens

This group contains 56 taxa and represents a miscellany of species preferring open disturbed ground and well-drained fresh soils. Such habitats are likely to have been much more widespread in cold stage stadial environments, where fluvial sediments may have been actively aggrading in river valleys, and periglacial processes such as solifluction and cryoturbation were interrupting soil-forming processes. The list includes species of a range of soils, but a large number presently prefer calcareous soils.

Species occurring in five or more samples include *Salix herbacea* (m42, p4), *S. polaris* (m6), *S. lapponum* (m5), *Minuartia verna* (m10), *Lychnis alpina* (m15), *Thalictrum alpinum* (m28), *T. minus* (m11), *Draba incana* (m37), *Saxifraga oppositifolia* (m20, p3), *Potentilla fruticosa* (m6), *P. crantzii* (m12), *Linum catharticum* (m8, p4), *Armeria maritima* (m72, p41), *Jasione montana* (p6), *Saussurea alpina* (p7), *Leontodon autumnalis* (m13) and *Allium schoenoprasum* (m17).

The majority of species belong to Preston and Hill's more northern categories, a contrast to species of the other vegetation types considered here

(Table 12.1). This clearly indicates that northern species were able to move south to join more temperate species, through selection of biotypes, climatic change, change in available habitats or change in ecological factors such as competition, soil conditions, etc.

Dunes

Most species in this group of 19 taxa also occur in other vegetation categories, especially grassland. They may have been present in communities on cover sand or on sandy fluvial sediments (e.g. Plates 4, 7B). Species represented in five or more samples have wider occurrences and they include *Salix repens* (m8), *Thalictrum minus* (m11), *Draba incana* (m37), *Potentilla anserina* (m83), *Linum catharticum* (m8, p4) and *Campanula rotundifolia* (m87). These species are mostly Boreo-temperate and wide-ranging to the east (Table 12.1).

Heaths

At the present day heathland communities are a major part of the shrub tundras of the north, where they give rise to particular pollen and macro assemblages (e.g. Lichti-Federovich & Ritchie 1968; West *et al.* 1993). The occurrence of such vegetation and its origin in cold stage times is therefore of special interest. The following discussion includes the taxa listed in the vegetation field of the PLANTHAB table under the categories H (heath), B (bog) and Mo (moor). Representation of Ericales pollen in the pollen samples is recorded in the PLSAMPLE table.

As described above, in the Devensian interstadials there is clear evidence from the pollen and macroscopic assemblages of the presence of heathland. Similarly, in north-west Europe, heathland is recorded in sections of Early Weichselian interstadial age (Behre 1989). In stadial floras there is a contrast between Early Pleistocene and Late Pleistocene floras (see Figures 8.8, 8.9, 8.10). The Early Pleistocene marine or estuarine cold stage sediments contain pollen taxa of the Ericales in substantial frequency. These have been interpreted as indicating the presence of oceanic heath in the earliest cold stages (Ba, Th) (Figure 12.3) (West 1980a, b). Assemblages with high frequencies of Ericales pollen have also been recorded in cold stages of the Early Pleistocene in north-west Europe (Menke 1975), indicating the wide development of heathland during these times.

In the Late Pleistocene stadials, however, the frequency of Ericales taxa is much lower. The majority of species associated with heath or other acidic habitats (categories H, B, Mo) have a very low frequency of occurrence and the great majority occur in a variety of vegetation types. The most recorded in samples include the following.

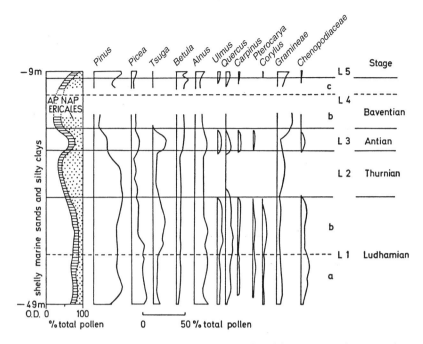

Figure 12.3. Pollen diagram from a long Early Pleistocene marine record at Ludham, Norfolk. The cold stages (Thurnian, Baventian) are marked by decreased tree pollen (AP) percentages and increased non-tree (NAP) pollen percentages, including Ericales pollen (West 1980b).

H taxa: *Potentilla palustris* (m25), *Rumex acetosella* (m25, p26), *Potentilla erecta* (m11), *Huperzia selago* (p20) and *Pedicularis palustris* (m9).

B taxa: *Menyanthes trifoliata* (m43, p1), *Eriophorum vaginatum* (m23).

Mo taxa: *Lycopodium annotinum* (p24), *Betula nana* (m54, p15), *Linum catharticum* (m8, p4).

There is no species which occurs consistently and frequently and indicates widespread heathland, rather a collection of species of varied habitat, many associated with grassland as well as heath, bogs or moors.

Examination of the pollen and macro records enlarges the view. The problem here is that freqencies of pollen (and macros) are usually very low (typically 1–2 per cent). The possibility of reworking, which is always present in cold stage assemblages, has to be borne in mind in the interpretation of these low percentages, especially when pollen grains of Ericales taxa, and *Sphagnum* spores, have thick and resistant walls.

Stadial pollen taxa of the Ericales include Ericales, Ericaceae,

Bruckenthalia, Calluna and *C. vulgaris, Rhododendron,* cf. *Loiseluria pro-cumbens* type, *Empetrum* and *E. nigrum* type. Apart from the Early Pleistocene, the percentages of these taxa and of *Sphagnum* are very low, except for higher percentages of Ericales pollen in the Early Devensian at Wing and the Early Wolstonian of Marks Tey, and of *Empetrum* in the Early Wolstonian of Marks Tey. There are also rather higher values in Wolstonian assemblages at Tottenhill, where sediments directly overlie Hoxnian sediments, and at Broome, where there is also reworked Hoxnian pollen.

At Wing, Hall (1980) considered that there was some evidence for the for-mation of ombrogenous peat, with *Empetrum, Vaccinium, Eriophorum vag-inatum* and ombrogenous mire indicators *Rubus chamaemorus* and species of moss at the end of the temperate Ipswichian Stage, but the predominant formation appeared to be grassland. At Marks Tey, Turner (1970) believed the *Empetrum* frequencies in the Early Wolstonian were largely reworked from the Late Hoxnian heathland, with the predominant vegetation again grassland. Evidence from these sites thus indicates the presence of peatland in the Early Devensian, more doubtfully in the Early Wolstonian, with grassland the major formation. Reworking of Ericales pollen taxa at the Tottenhill and Broome sites is more clearly demonstrated.

The low frequency of these pollen records is paralleled by the low number of macroscopic records of Ericales taxa. Ericaceae seeds (6) have been recorded from the Middle Devensian (Brimpton) and Wolstonian (1, Somersham), *Erica* cf. *tetralix* (m1) from the Middle Devensian (Brimpton), *Bruckenthalia spiculifolia* (many) from the Early Devensian (Beetley), *Calluna vulgaris* (m1) from the Middle Devensian (Sandy), *Arctostaphylos uva-ursi* (present) from the Middle/Late Devensian (Lea Valley), *Vaccinium* (usually single finds in samples) from Devensian sites, *V. oxycoccus* (m1) from the Middle Devensian (Beetley) and several cf. iden-tifications from the Early Devensian (Wing), *V. myrtillus* from the Devensian (cf. m2, Block Fen) and Wolstonian (m1, Broome), *Empetrum* and *E. nigrum* rarely from several Devensian sites, Early, Middle and Late.

Taking the pollen and macro records as a whole, there is little evidence for extensive heathland in Late Pleistocene stadial vegetation in Britain. The low frequencies of pollen of Ericales taxa and of *Sphagnum* spores, and the thin record for macros of heath species indicates persistence of local communities with these taxa throughout the cold stages, if we dis-count total reworking as an explanation for their presence. From such com-munities expansion of the species concerned was possible in the interstadials. But there is no evidence for widespread stadial vegetation

similar to shrub tundra, as there is no evidence for widespread peat formation often associated with shrub tundra at the present time.

Trees, shrubs

In the stadials, *Juniperus*, *Betula* and *Salix* are represented by macroscopic
remains and pollen. Tree birch fruits are recorded macroscopically much
less frequently than dwarf birch; the reverse is so in the pollen record. Both
shrub and dwarf willows are recorded macroscopically. The contribution to
the pollen record by these two genera is very greatly overshadowed by the
herbaceous pollen percentages, indicating local occurrence limited to more
favourable growth conditions of moisture and shelter, such as might be
found in floodplains in valleys, where there are also favourable conditions
for preservation. At Somersham, site SAG, unusually high frequencies of
Salix pollen with macroscopic remains of *Salix* were recorded in such a situation. The representation of tree/shrub pollen in the pollen samples is
recorded in the PLSAMPLE table.

A parallel is seen in the *Salix* shrub communities occurring in sheltered
valleys in tundra of the low Arctic or to valley communities with tree
birches and many species of *Salix* (including *S. viminalis*) in the Eurasian
steppe (e.g. Printz 1921). While the tree birches and shrub willows (e.g. *S.
phylicifolia*, *S. viminalis*) may have been confined to such sheltered sites, the
dwarf birch and willows (*Betula nana*, *Salix reticulata*, *S. herbacea*, *S.
polaris*) are also likely to have been more widespread in locally favourable
situations.

The reduction of tree cover in the stadials is discussed from the climate
point of view in Chapter 13.

Woodland, scrub

Forty-nine taxa are in this group, 31 occurring in other vegetation categories. The tree species in the list are recorded mostly in interstadial samples.
Other records of trees, such as *Abies alba*, *Pinus sylvestris*, *Cornus sanguinea* and *Alnus glutinosa* may be reworked. Further genera, including
Juniperus, *Betula* and *Salix* are better represented, as discussed above. The
much less common *Bruckenthalia spiculifolia* (m5) may be added to these
genera.

All the herb species represented in five or more samples have wider preferences. They include *Urtica dioica* (m48), *Rumex acetosa* (m8, p26),
Stellaria graminea (m7), *Caltha palustris* (m14), *Filipendula ulmaria* (m17,
p5) and *Succisa pratensis* (p13). None of these are strictly shade plants,
but they, and the less frequently recorded species in the group, may be

associated with willow or birch scrub communities on floodplains and elsewhere.

Most of the taxa in the group are Boreo-temperate and wide-ranging to the east (Table 12.1).

Fens, flushes, marshes

The species here are labelled Fe (fens, flushes) and Ma (marshes) in the vegetation field of the PLANTHAB table, with 49 taxa in the Fe category and 58 in the Ma category. The large number of taxa reflects the good preservation potential of species of the communities in the organic sediments deposited in these environments, and indicates rich communities such as are not presently found in high latitudes.

Most species are common to both categories and many prefer eutrophic conditions. The marsh species include *Eleocharis palustris* (m62), but many are maritime or prefer saline soils. These include *Suaeda maritima* (m5), *Glaux maritima* (m19), *Armeria maritima* (m72, p41) and *Juncus gerardii* (m6). The shrubs *Salix repens* (Ma, m8) and *S. viminalis* (Ma & Fe, m7, p1) are recorded in low frequency. The more frequent remaining herbs in both categories include *Urtica dioica* (m48), *Stellaria palustris* (m10), *Caltha palustris* (m14), *Ranunculus lingua* (m16), *Thalictrum flavum* (m38), *Parnassia palustris* (m9), *Filipendula ulmaria* (m17), *Potentilla palustris* (m25), *P. erecta* (m11), *Hydrocotyle vulgaris* (m6), *Berula erecta* (m8), *Menyanthes trifoliata* (m43, p1), *Lycopus europaeus* (m25), *Valeriana dioica* (m3, p2), *Succisa pratensis* (p13) and *Typha latifolia* (m7, p76).

The assemblage of fen and marsh plants is typically temperate in present distribution. All except one species with a frequency of occurring in five or more samples are in the Boreo-temperate, Temperate and Southern-temperate categories of Preston & Hill (1997). The exception is *Caltha palustris* of the circumpolar Wide-boreal category. The more southern species include *Salix viminalis, Suaeda maritima, Hydrocotyle vulgaris, Berula erecta, Lycopus europaeus* and *Typha latifolia*.

Aquatic vegetation

A rich aquatic flora is present in the stadial periods, recorded by the taxa labelled A in the vegetation field of the PLANTHAB table. Preferences relating to water are in the water field of the same table. Numbers of taxa in the field categories are given in Table 10.1.

The species are predominantly those preferring meso- to eutrophic conditions. This is presumably a reflection not only of the fact that most sites are in the softer rock areas of south and east Britain, but also that water bodies have been subject to enrichment by periglacial processes such as solifluction. The few preferring oligo- to mesotrophic waters (e.g. *Montia fontana* (m4), *Potamogeton polygonifolius* (m1)) are rarely recorded, except for *Myriophyllum alterniflorum* (m16, p19) and *Potamogeton alpinus* (m28).

Few species have a preference for deeper waters. They include *Nuphar lutea* (m9), and *Myriophyllum verticillatum* (m5, p46), *M. spicatum* (m33, p46) and *M. alterniflorum* (m16, p19). Most species prefer shallow water bodies or wet habitats marginal to water bodies. The more commonly recorded shallower water species include *Ceratophyllum demersum* (m10), *Ranunculus hyperboreus* (m28), *R. sceleratus* (m58), *Hippuris vulgaris* (m97), *Menyanthes trifoliata* (m 43, p1), *Callitriche* spp. (m12), *Sagittaria sagittifolia* (m9), *Alisma plantago-aquatica* (m27), the southern species *Damasonium alisma* (m4), *Zannichellia palustris* (m57) and *Sparganium minimum* (m5). Emergent and marginal aquatics also include *Ranunculus flammula* (m34), *Berula erecta* (m8) and *Cicuta virosa* (m7).

A remarkable number of taxa of Potamogetonaceae (22), occurring in shallow and deeper water, have been identified. They are all presently British except for *Potamogeton vaginatus;* their distribution has been described by Preston (1995). The most recorded are the floating leaf species *P. natans* (m28), *P. gramineus* (m18), *P. alpinus* (m28), and the submerged species *P. praelongus* (m24), *P. pusillus* (m9), *P. obtusifolius* (m9), *P. berchtoldii* (m13), *P. acutifolius* (m5), *P. crispus* (m30), *P. filiformis* (m80), *P. vaginatus* (m13), *P. pectinatus* (m28) and *Groenlandia densa* (m53). This range of species is far greater than occurs in the present Arctic. A number have a more southern distribution (see Bell 1969): *P. acutifolius*, *P. crispus*, *Groenlandia densa*. The most frequently identified species, *P. filiformis*, is a circumpolar species of the low Arctic. The species has a present northern distribution in Britain, which does not overlap with the south-eastern *P. acutifolius*.

Of the emergent shallow water or reedswamp aquatics, the most commonly recorded are *Sparganium erectum* (m27), *Scirpus lacustris* (m64) and *Eleocharis palustris* (m62).

It is not unexpected that the aquatic flora, with strong preservation potential, has such an excellent record of species. Most have a wide circumboreal distribution at the present day (Table 12.1), and, like the terrestrial flora, show a mixture of distribution types, with the number of species an indication of a favourable climate. The more northern species are accompanied by the southern species *Potamogeton pusillus*, *P. crispus* and *Zannichellia palustris*.

Plants associated with salinity

Saline habitats (Plate 7A) and their origin are discussed in Chapter 10. Table 11.2 lists Bell's (1969) cold stage halophytes. The PLANTHAB table lists taxa with some preference for a degree of salinity in their habitats. Three categories are used in the salinity field: brackish, maritime and saline. A small number of species which occur in brackish water are recorded: *Ranunculus hyperboreus* (m28), *Potamogeton vaginatus* (m13), *P. filiformis* (m80), *P. pusillus* (m9), *Zannichellia palustris* (m57), *Naias marina* (m2), *Scirpus lacustris* ssp. *tabernaemontani* (m3). Fewer species, usually found fossil more rarely, are associated with saline conditions in marsh or grassland. They include *Chenopodium rubrum* (m3), *Stellaria media* (m12), *Glaux maritima* (m19), *Plantago maritima* (m5, p120), *Triglochin maritima* (m5) and *Juncus gerardii* (m6). A greater number of species, over 30, have a maritime or coastal distribution, of which the most commonly recorded are *Atriplex hastata* (m6), *Ranunculus flammula* (m34), *Glaux maritima* (m19), *Armeria maritima* (m72, p41), *Silene vulgaris* ssp. *maritima* (m9), *Plantago coronopus* (p15), *P. maritima* (m5, p120) and *Juncus gerardii* (m6).

In addition to these macro and pollen records, pollen frequencies of *Plantago maritima* in some stadial samples reach nearly 20 per cent total pollen (e.g. Block Fen). Evidently there were opportunities for plants now behaving as halophytes or able to withstand brackish conditions to have a much wider inland distribution in cold stages, either for reasons of reduced competition or because of salinity induced by climate and soil conditions (Chapter 10).

Plants of man-made habitats

Many species in the PLANTHAB table occur in man-made or semi-natural habitats as well as in natural communities. There are records of 81 species which are nowadays associated with man-made or semi-natural habitats. These are listed in the PLANTHAB table in the places field under the categories of plants of cultivated ground (Cg), disturbed ground (Ds), fields (Fi), road/waysides (Ro), and waste places (Wa). They include plants now classed as weeds or as ruderals (plants growing in waste places). This large number of species reflects the variety of open and disturbed habitats, of varying soil type and drainage conditions, available in cold stage environments, especially in areas where sediment deposition is active from time to time and place to place, as on floodplains. Such habitats were also present in areas of solifluction, mass flow and cryoturbation. It has been suggested that trampling by large mammals on floodplains also engendered habitats suitable for some of the species recorded (e.g. *Coronopus squamatus*).

Twenty-five of the species are recorded in five or more samples. These include *Polygonum aviculare* (m11, p34), *P. lapathifolium* (m6), *Bilderdykia convolvulus* (m1, p8), *Rumex acetosella* (m25, p26), *Atriplex hastata* (m6), *Montia fontana* (m5, p4), *Stellaria media* (m12), *Cerastium arvense* (m29), *Ranunculus repens* (m28), *Capsella bursa-pastoris* (m5), *Diplotaxis tenuifolia* (m41), *Potentilla anserina* (m83), *Viola tricolor* (m6), *Glaux maritima* (m19), *Stachys palustris* (m5), *Plantago major* (m7), *Achillea millefolium* (m12), *Matricaria perforata* (m5), *Cirsium arvense* (m6), *Centaurea nigra* (p20), *Leontodon autumnalis* (m13), *Sonchus oleraceus* (m9), *Taraxacum officinale* agg. (m13) and *Juncus bufonius* (m9).

The versatility of these species is demonstrated by the number of them (20) which are described by Preston & Hill (1997) as widely naturalised in the British Isles. In terms of present distribution of the 25 species, most species are in the Boreo-temperate and Wide-temperate categories of Preston & Hill (1997), and in the categories wide-ranging to the east (Table 12.1). None is in the Boreal-montane and more northern categories and only two are in the Southern-temperate category.

The stadial flora offers a great contrast with the present Arctic flora in terms of the distributions mentioned above, as well as in the larger number of annuals (Chapter 9). But just as in the Arctic, where a deep active layer in well-drained soils on gravel and sands promotes a warmth not found in wetter tundra soils even in the short summer, similar conditions but with longer summers at much more southerly latitudes in the stadial periods must have produced favourable conditions for plant growth. The plant communities present are likely to have nourished a fauna unlike that found in the Arctic today.

Comments on sites and local vegetation

Most stadial sediments yielding floras are situated in floodplain gravel sequences. The richness of these floras in macroscopic remains of herbs and shrubs (e.g. at Earith, Somersham, Wretton) must reflect the variety of habitats available on floodplains, in terms of the range of sediments available for colonisation and their water-holding capacity. The floras are primarily of local origin, contrasting with the wider and regional vegetation indicated by the pollen assemblages.

For example, the rich macroscopic flora (71 taxa) at the Devensian Wretton site WUB occurred in a silty sand where floodplain sands abutted Gault Clay at the margin of the floodplain. Many dry, open ground, species were identified, in contrast to fewer such species in more organic neighbouring sediments (site WU), a difference resulting from taphonomy. The analyses of many sites in the Devensian floodplain at Wretton showed the variety of assemblages possible, with both macro and pollen evidence for particular taxa, including *Armeria maritima*, *Campanula rotundifolia*, *Ranunculus acris* and *Sparganium erectum*. The variety of herb pollen taxa at Wretton (site WH) is shown by the Early Devensian pollen diagram of Figure 12.2.

The more organic sediments of pools on floodplains, likely to be formed in abandoned channels, have generally poorer macroscopic floras, with few aquatic species, including for example *Potamogeton* spp., *Ranunculus* subg. *Batrachium* and *Zannichellia palustris*. Pollen diagrams from such sites, e.g. the Devensian Beetley site RR, show high frequencies of Cyperaceae pollen, originating in local sedge communities on the floodplain. But there is also pollen of dry land plants such as *Armeria*, *Campanula*, *Helianthemum* and *Plantago* from the floodplain or neighbouring valley slopes.

Rarely, vegetational changes are revealed by pollen diagrams from pool sediments, as at the Middle Devensian Beetley site WW, where changes through 50 cm of sediment were recorded. Here the basal pollen spectra in silty mud show high frequencies of pollen of Gramineae and *Armeria*, with *Campanula* type, *Plantago media/major* and *Polygonum bistorta/viviparum*. At higher levels, in more organic sediment, Gramineae pollen decreased and Cyperaceae and *Ranunculus* cf. *aquatilis* pollen increased, with abundant achenes of *Ranunculus* subg. *Batrachium* in the upper part of the pool filling in a nearby section (Beetley XX). The lower pollen spectra appear to represent a herb-rich grassland, pollen content changing as a hydrosere evolved.

A small number of sections show organic sediments of small pools associated with solifluction sediments on valley slopes. Two such sections are the Devensian sites R and W at Beetley. Both show very high frequencies of Gramineae pollen and low frequencies of Cyperaceae pollen, taken to indicate regional grassland. At the top of the 20 cm-thick R section, Cyperaceae pollen increases, as in the WW section described above, indicating hydroseral successsion; abundant macroscopic remains of *Carex*, *Juncus*, *Luzula* and *Viola* indicate the nature of the flora local to the pool within the solifluction sheet.

Vegetation at particular times

The variability of stadial macroscopic and pollen assemblages, resulting from the complexities of the vegetation and the taphonomy, make for a real difficulty in distinguishing vegetation variation over time at the cold stage sites being considered. This is not such a problem in the interpretation of pollen diagrams from long lake sequences, such as that at Grande Pile (Woillard 1979), where regional vegetation change can express itself more clearly. On the other hand, the fossil assemblages of such lake sediments give much less information on species presence and taphonomy.

However, there are certain pollen assemblages which recur at times in cold stages. These indicate the recurrence of a particular vegetation type, which may be controlled by climatic change or by more local changes of plant communities and taphonomy. A combination which appears in pollen spectra at several times is that of notable frequencies of *Artemisia* and *Plantago maritima*. This combination is recorded in certain spectra at Block Fen (De, 1 De), Brandon (e Wo), Somersham (1 De) and Wretton (e De). At Brandon (Maddy *et al.* 1994), it was suggested that a continental climate produced saline soil conditions, an indication supported by the presence of a salt-marsh beetle.

At Block Fen and Wretton, other cold stage pollen spectra do not show this combination. Either there were times of marked continental climate or local environmental conditions favoured the combination. In respect of the latter, Pennington (1986) suggested that *Artemisia* pollen was more abundant in the Devensian late-glacial where the sites studied were on outwash sands and gravels. All the sites mentioned above are associated with sand and gravel sequences, but Block Fen and Wretton also show spectra without the combination. There is therefore the possibility that this combination represents periods of marked continentality of climate. Certainly, both *Artemisia* and *Plantago maritima* are plants of saline soils in the steppe. Even so, *Artemisia* is rare in pollen spectra of the Early Devensian site at Wing, within an area of Lias clay, and the effects of local environmental factors cannot be discounted.

Pollen spectra with notable frequencies of *Artemisia* have also been recorded at an early-glacial site (Marks Tey III, zone e Gi 1) and at several late-glacial times, as follows: in the Late Beestonian (1 Be substage a) at West Runton, in Late Wolstonian at Ilford (zone b), Selsey (zone b) and Bobbitshole (Ipswich) (zone b). Low frequencies of pollen of *Plantago maritima* occur in these spectra at Selsey and Ilford. At Bobbitshole the zone b assembage is underlain by loess. These records suggest that continentality of climate was a feature of these late-glacial periods. This is a contrast with the Anglian late-glacial, where pollen of *Hippophae* (and macroscopic remains of *H. rhamnoides*) reaches high frequencies in lake sediments immediatialy preceding the temperate Hoxnian Stage and succeeding in some sites Anglian glacigenic sediments. *Hippophae* pollen is rarely recorded in other cold stage sediments.

Helianthemum pollen and *H. canum* are often found in the assemblages with notable frequencies of *Artemisia* pollen, for example, in Early Devensian at Wretton and the Late Devensian at Somersham. *H. canum* is in Preston & Hill's (1997) Mediterranean-montane floristic element, associated with a southern distribution of some warmth of summer and open, well-drained habitats.

Notes on particular taxa

The following comments concern a number of taxa which are of interest in considering stadial vegetation and flora. Many of these, and other taxa, are considered in terms of ecology or identification by Dickson (1970) and Godwin (1975), as well as in the descriptions of stadial assemblages, especially those by Bell (1968, 1969, 1970), Lambert *et al.* (1963) and West *et al.* (1974). Occurrences of all the taxa are available in the database tables. The floristic elements mentioned are those of Preston & Hill (1997).

Allium schoenoprasum is a circumpolar Boreo-arctic Montane species of damp and grassy places on river banks and lake shores (Plate 8B). The occurrence in stadial assemblages is therefore not surprising. There are many records from Devensian sites and from one Cromer Complex cold stage site (Ardleigh).

Androsace septentrionalis, a non-British species, is an annual of dry sandy places, with a north and central distribution in Europe. It has been recorded in the Early Wolstonian (Brandon) and the Middle Devensian (Upton Warren). The habitat requirements are similar to *Corispermum* (see below).

Arenaria ciliata. Seeds of *Arenaria* recovered from the Middle Devensian site at Earith were studied by Bell (1970), who identified many as the European Arctic-montane *A. ciliata*. Bell considered that the morphology of these bore a close resemblance to those of the present Irish population rather than those from other European and Greenland populations. The species has also been identified in other Middle Devensian assemblages. The habitat preferred is dry and gravelly, a likely habitat on stadial flood-plains.

Armeria. The polymorphic *A. maritima*, a Circumpolar Wide-boreal species of arctic and temperate regions, occurs frequently in stadial floras from the Beestonian onwards. Pollen of *Armeria* is even more frequently found, with both A and B types. The species was clearly widespread and successful in open communities of the floodplains and elsewhere. The variability and wide range of *Armeria* make the plant an unsuitable measure of past climate.

Betula nana, identified from fruits, leaves and more rarely pollen, has been commonly recorded in stadial floras from the Beestonian onwards. At present it has a Circumpolar Boreo-arctic Montane distribution reaching south into temperate areas. The macroscopic remains are rarely found in abundance, and it is possible that the shrub was confined to damper sheltered areas, rather than in widespread moor-like or peaty habitats.

Campanula rotundifolia, a variable Circumpolar Boreo-temperate species, has been commonly identified in many stadial floras of Devensian, Wolstonian, Middle Anglian and Cromer Complex cold stage age. The species clearly has a wide climatic tolerance, and is presently in arctic to temperate regions, preferring dry, open or grassy habitats, such as would be present in stadial times.

Corispermum. Fruits of this non-British central European genus have been recorded from a number of Early, Middle and Late Devensian sites and fron a Wolstonian site. The many species favour dry sandy and gravelly places. The records include *C.* cf. *hyssopifolium*, which has an eastern European distribution.

Damasonium alisma. Seeds of this Mediterranean-Atlantic species have been recorded in three Middle and Late Devensian sites. *D. alisma* has a remarkably southern distribution at present. It is very variable in form (see e.g. Glück 1936) and favours intermittently wet habitats. These may be related to periods of wetness in the spring and early summer, associated with thaw and a continental climate. The habitat requirement has similarities with that of *Ranunculus hyperboreus*, which has a much more northern distribution today (see below).

Dianthus. Seeds of this complex genus have been recorded at several Devensian sites and one Wolstonian site. Seeds have been identified to the genus, or as *D.* cf. *gratianopolis*, *D. deltoides* and *D.* cf. *carthusianorum*. These are presently plants of dry grassland, rocks or woods, often on basic soils; the former are likely stadial habitats.

Diplotaxis tenuifolia has a European Temperate distribution and is considered doubtfully native in the British Isles. Although with this southern type of distribution, the species has very numerous records in the Devensian, with a few in the Wolstonian and Cromer Complex cold stages. The seeds are sometimes found in large quantities, probably a taphonomic effect of their buoyancy, which is described by Bell (1970). The present habitat includes fields, waysides and waste places, with a stadial equivalent of floodplains and disturbed ground.

Draba incana has a European Boreo-arctic Montane distribution, with a northern distribution in Britain, from sea level to mountains, in dunes to rocky areas. Seeds and fruit valves have been recorded frequently in the Devensian and more rarely in the Wolstonian. Stadial habitats similar to those favoured today must have been frequent on floodplains and elsewhere.

Dryas octopetala has easily recognisable leaves which have been frequently found on the continent in Weichselian (Devensian) late-glacial sediments, giving rise to the 'Older and Younger Dryas' terminology of this period. However, in the stadial record considered here, leaves and pollen are remarkably rare, with a few Late Devensian records only. There are considerably more records from the Devensian late-glacial *sensu stricto*. *D. octopetala* has a Circumpolar Arctic-montane distribution, and in the British Isles is found on basic rocks in montane areas, but descending to near sea level in Co. Clare and Sutherland. The climatic range within the distribution area is thus wide, soil conditions and competition evidently playing a more important role. It is unexpected that in many assemblages of stadial 'leaf floras' *Dryas* is absent, though basic soil conditions appear to be present.

Euphorbia cyparissias. Seeds of this European Temperate species have been recovered from a few Middle and Late Devensian sites and a Cromer Complex cold stage site (Ardleigh). It is a plant of calcareous grassland in Britain, considered to be introduced. More widely in Europe it favours also waysides and cultivated ground, and so falls into a category of such plants in the stadial flora.

Helianthemum canum is one of the most southern species in the stadial flora, having a Mediterranean-montane distribution. In the British Isles it is now a rare plant of open rocky limestone pastures. Capsules, leaves and seeds have been recognised in many Devensian assemblages and at a Wolstonian site. *Helianthemum* pollen has also been identified at a number of Devensian sites. A remarkable occurrence is in the Devensian assemblage at Somersham (Some SAG), where macroscopic remains of *H. canum* were found, together with high frequencies of *Helianthemum* pollen, in sediments with a radiocarbon date of 18,750 years BP. The site is within 60 km of the Late Devensian ice margin to the north at about the same time. *H. canum* is a very variable species (Griffiths & Proctor 1956), but generally appears to demand good summer temperatures in its main range today, though still distributed rarely in more northern and cooler areas.

Herniaria has been recorded in Early, Middle and Late Devensian and Late Wolstonian assemblages. *H. glabra* has been recorded in a Middle Devensian and an Early Wolstonian site. This Eurosiberian Temperate species, like many stadial species, prefers dry sandy habitats.

Hippophae. Pollen of *Hippophae* has been recorded principally in the Late Anglian in Britain, where relatively high frequencies are accompanied by leaf scales of *H. rhamnoides*. Similar pollen frequencies are found in the Pre-Gortian in Ireland. Otherwise, the few pollen records (of lower frequency) occur in the Late Beestonian at West Runton and the Late Devensian at Colney Heath. *Hippophae* appears in Britain as a successful and rapid coloniser of fresh soils exposed by the retreat of Anglian ice, perhaps favoured by climatic amelioration towards the end of the cold stage, as it appears to have been at the end of the Devensian. But the occurrence in the Late Anglian, unique in the record, suggests that a combination of chance and climate determined its spread and abundance.

Koenigia islandica is a species of open and fresh soils and has a Circumpolar Arctic-montane distribution. It is now confined in Britain to Skye and Mull. There are four Middle or Early-Middle Devensian pollen records from the Scottish sites of Burn of Benholm and Sourlie. Absence of records further south may suggest stadial refuges in north Britain.

Linum perenne agg. is a complex species with a Circumpolar Boreo-temperate distribution. *L. anglicum*, which has been identified ina number of stadial samples, is included in the aggregate species. *L. perenne* occurs in the arctic and temperate regions, as a pioneer of open ground and a constituent of grassland on basic soils. The seeds have been recorded frequently in Devensian assemblages, and also in those of Wolstonian, Anglian and

Cromer Complex cold stages. The species was evidently a widespread stadial plant.

Lychnis alpina. Seeds of this European Boreo-arctic Montane species have been identified in Devensian, Middle and Late Devensian assemblages and in a Wolstonian assemblage. The Earith fossils were described by Bell (1968), who considered they had a cell-shape similar to Scandinavian populations. In the Arctic, *L. alpina* prefers damp sandy or gravelly habitats, such as would be found on stadial floodplains. Today it is a very rare alpine plant in Britain.

Medicago sativa ssp. *falcata* has a Eurosiberian Southern-temperate distribution. The subspecies is thought to be native in East Anglia, where it grows in grassy areas on dry gravelly soils. Pods and seeds have been recorded in Middle and Late Devensian assemblages from gravel floodplains, as with *Onobrychis viciifolia.*

Minuartia verna, M. rubella. M. verna is a Eurasian Boreal-montane plant, absent from Scandinavia. It is locally distributed in Britain, restricted to base-rich rocky areas and pastures mainly in northern England. Seeds have been recorded from a number of Middle and Late Devensian sites and from the Cromer Complex cold stage site at Ardleigh. The distinction between the seeds of this species and those of *M. rubella* are discussed by Bell (1970). The latter is a Circumboreal Arctic-montane plant, also recorded from a few Middle Devensian sites, and now rare on base-rich rocks on Scottish mountains.

Onobrychis viciifolia is a Eurosiberian Temperate species, regarded as one of the southern elements in the stadial flora. It has been considered doubtfully native in calcareous grasslands and open ground in south-east Britain. There are a number of Devensian stadial records of pods and seeds, indicating that the species was certainly present over a long period of time in the Middle and Late Devensian, in grassland and more open habitats.

Papaver. Considering the present widespread occurrence and abundance of *Papaver* spp. in the arctic and the presence of apparently favourable sites on stadial floodplains, it might be expected that the genus would be well represented in the stadial flora. Records, however, are not common. *Papaver* sect. *Scapiflora* seeds have been recorded rarely in the Devensian (Middle and Late), Wolstonian and Cromer Complex cold stages. A number of seeds of

P. radicatum s.l. were identified at the Middle Devensian site at Sourlie, and rare cf. *Papaver* seeds at a Wolstonian and a Middle Devensian site.

Plantago. The five native species of *Plantago* have been identified in stadial assemblages from macroscopic remains and/or pollen. The taxa recorded most frequently are *P. media/major* and *P. maritima*, with records in cold stages back to the Pre-Pastonian. *P. lanceolata* has records in the Early, Middle and Late Devensian and Wolstonian. *P. maritima* is the only one of these species widespread in the arctic, though *P. major* and *P. coronopus* are reported to be established in some arctic areas. The latter and *P. lanceolata* have a European Southern-temperate distribution, *P. media* a Eurasian Temperate distribution, *P. major* a Eurasian Wide-temperate distribution and the species reaching north the most, *P. maritima*, a Eurosiberian Wide-boreal distribution. All species evidently found suitable habitats of open and disturbed ground or in grassland during the stadia.

Polygonum aviculare agg. is a Circumpolar Wide-temperate species, considered to be introduced where it is now found in the arctic. But it has been recorded in many stadial floras back to the Cromer Complex cold stage (a cf. identification). At present *P. aviculare* is a plant of waysides, waste places and coasts. But it appears to have been a consistent member of stadial floras, where it would certainly find the open and disturbed habitats preferred.

Potentilla anserina has a Circumpolar Boreo-temperate distribution, hardly reaching the arctic, but extending from the coasts eastwards to the steppe. *P. anserina* is a plant of waysides, waste places, pastures, dunes and the coastal areas. It is one of the most common species of stadial assemblages, with abundant records in the Devensian and in earlier cold stages back to the Pre-Pastonian. Also, the species is tolerant of saline soils, which may have favoured its abundance.

Potentilla fruticosa. This shrub has a Circumpolar Boreal-montane distribution, scarcely reaching the arctic, and preferring damper and base-rich conditions in rocky places and river banks. *P. fruticosa* has a very disjunct distribution in western Europe, but eastwards it is more widespread, as in river valleys of southern Siberia and in North America. Seeds (and with a calyx at one site) have been recorded at a few sites in the Devensian and Wolstonian. At Somersham (Some SAG) seeds were recorded with those of *Helianthemum canum* in the Late Devensian near the margin of the ice (see above).

Ranunculus hyperboreus. A non-British species which has a present circumpolar arctic-montane distribution to high latitudes, *R. hyperboreus* is a plant of shallow freshwater or somewhat brackish pools, and damp open or mossy ground. There are many records of achenes from the Early, Middle and Late Devensian, and a few from the Wolstonian, Anglian and Pre-Pastonian. The habitat is one which would favour the preservation of the achenes. An abundance of achenes has been found in a few samples, e.g. in an Early Devensian sample at Wing, 245 achenes were found in a sample of 298 macros, with only one or two in samples above and below in similar sediment, possibly variation as a result of very local abundance of the plant.

Salix herbacea has a European Arctic-montane distribution, and is locally common in the mountains of northern Britain. The open habitats favoured have fresh and basic soils, such as those subject to solifluction, and the plant is particularly associated with late snow patches. *S. herbacea* leaves have been recorded at many sites, including those of Early, Middle and Late Devensian and Wolstonian age, but apparently not earlier than the Late Anglian. They also occur in assemblages thought to be of Devensian Upton Warren Interstadial Complex age. The deciduous leaves can be concentrated by transport processes in streams, and the result is seen in the numbers of leaves recorded in some assemblages (e.g. Beetley BB1, Late Devensian).

S. polaris. This non-British species has a more northern and less complete circumpolar distribution than *S. herbacea*. It favours open moist habitats, where snow has lain through the winter. The distinction between the leaves of the two species is described by Bell (1970). There are far fewer records for *S. polaris* than *S. herbacea*; they are from the Early, Middle (including the Upton Warren Interstadial) and Late Devensian and the Early Wolstonian. There are cf. identifications from Wolstonian, Early Anglian and Beestonian assemblages.

S. reticulata is a Circumpolar Arctic-montane species, now in Britain only in a limited number of Scottish mountain localities. It favours calcareous open soils, with snow lie in the northern part of the range. The very few records are from the Late Devensian and Wolstonian, with cf. records from the Late Devensian and Wolstonian. The Wolstonian record is of a large number of leaves from a pond sediment immediately extra-marginal to ice (Tottenhill TH AA).

Of these three low and creeping shrubs, *S. herbacea*, the most widespread in the British Isles today, has been recorded fossil the most widely. The record means that the plant was not uncommon on floodplains and in adjacent areas during the stadials of the later cold stages. It is likely that solifluction and snow patch areas in declivities of the regional landscape harboured the plant, with transport of the leaves to pools on floodplains. The other two species are much more rarely recorded, and must have found fewer suitable localities. *S. repens*, with a more southern distribution (Eurosiberian Boreo-temperate), also occurs rarely in stadial floras, as does the even more southerly *S. viminalis* (Eurasian Temperate). The combination of *Salix* species of differing present geographical range is striking, which must reflect the range of habitats and competition available as well as the climatic conditions and biotype content of the species.

Saxifraga oppositifolia (Plate 8A) has a Circumpolar Arctic-montane distribution and is a variable species with a wide range of open habitats, from dry polar desert to wet rocks and gravels. In Britain it is now very largely confined to northern mountains. There are numerous records from the Middle and Late Devensian, Wolstonian, Early Anglian and Beestonian, indicating the species is a long-standing member of the stadial floras.

Selaginella selaginoides has a Circumpolar Boreal-montane distribution, and is a plant of wide open habitat range, in damp grassland and moss-rich communities in northern Britain and Ireland, particularly in mountain areas. The megaspores have been commonly recorded in all parts of the Devensian, in the Wolstonian, Anglian, Beestonian and Pre-Pastonian. The megaspore record is supported by the microspore record of *Selaginella*. The megaspores often occur in large numbers in assemblages, perhaps concentrated by taphonomy.

Stellaria crassifolia. This non-British species, with a wide distribution in Scandinavia, has a circumpolar low-arctic distribution, in damp grassy areas, inland and coastal. It has been recorded rarely in the Middle Devensian, but with more cf. identifications in the Middle and Late Devensian and in the Cromer Complex cold stage.

Urtica dioica has a Eurosiberian Boreo-temperate distribution, but reaches the arctic in eastern Europe and elsewhere. *U. dioica* is common in temperate regions of Europe, occurring especially where soils are enriched near habitations, but it also occurs in damp woods and tall-herb communities. The achenes have been identified in many Early, Middle and Late

Devensian assemblages, and in the earlier cold stage floras, indicating a long history in stadial floras.

Viola. The record of this genus in cold stages extends back to the Cromer Complex cold stage. The seeds are commonly found in stadial assemblages, but species can rarely be identified. *V. palustris* and *V. tricolor* have been identified in a small number of sites, mainly Devensian. A number of British species are associated with pastures and damp grassy areas or with cultivated or waste ground (e.g. *V. tricolor*, which can rapidly colonise gravelly areas), habitats which were present in the stadials.

13

Evidence of climate

One of the main aims of studying the cold stage flora is the reconstruction
of the climates. Interpretation of the climates of cold stages from the plant
fossil record is fraught with difficulty, making the subject a particular challenge. Any interpretation has to take into account the geological evidence
for climate, as well as the biological evidence. The latter involves taxonomy,
phytogeography, biology of the species and taphonomy; in all these areas
our knowledge needs enlargement.

The recognition of 'glacial' plants in Quaternary sediments during the
late nineteenth century led to the realisation that northern plants extended
south during cold periods. This also suggested a simple supposition of
'glacial' climates in present temperate areas, which accompanied the glaciations of the Quaternary. The discovery of the oxygen isotope record of
marine sediments (e.g. Emiliani 1955; Shackleton & Opdyke 1973) showed
an expression of continuous climatic change in the Quaternary. Variation
in the δ ^{18}O curve (e.g. in the Last Cold Stage, Figure 13.1A) is associated
mainly with global glacierisation, with enrichment in the oceans during
glacial times and depletion at times of reduced glacierisation. The curve
thus shows times of global ice increase, occupying major periods of time,
alternating with shorter times of ice decrease. The former are taken to represent glacial times (cold stages, cryomers) and the latter interglacial times
(temperate stages, thermomers). Webb (1988) pointed out that oxygen isotopes provide a univariate measure of the multivariate entity which is
climate, and that the exact mixture of climatic variables that cause oxygen
isotopes to change may not match those required for the vegetation to
change. The simplicity of the oxygen isotope curve masks the complexity of
terrestrial climates and terrestrial vegetation. This complexity has to be
superimposed on the oxygen isotope curve, which forms a base for consideration of the gross stratigraphy.

227

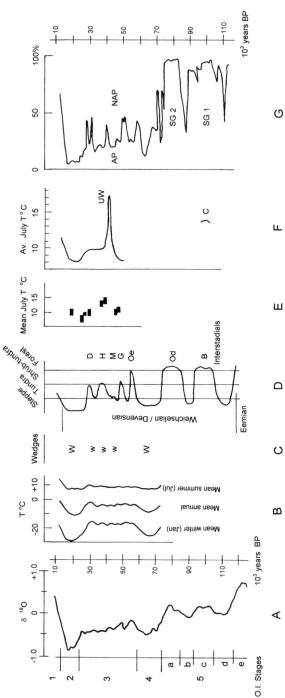

Figure 13.1. Evidence relating to climatic change during the Last Cold Stage (Devensian, Weichselian).
A. Oxygen Isotope Stages and radiocarbon chronology. After Martinson (1987) and Behre & van der Plicht (1992).
B. Temperature changes in the Netherlands Pleniglacial, based on periglacial phenomena and flora. After Vandenberghe (1992a).
C. Periglacial wedge structures in the Netherlands Pleniglacial. W, major ice wedge casts; w, minor wedges. After Vandenberghe (1992a).
D. Changes in Last Cold Stage vegetation in northern central Europe. Interstadials: B, Brørup; Od, Odderade; Oe, Oerel; G, Glinde; M, Moershoofd; H, Hengelo; D, Denekamp. After Behre (1989).
E. Estimations of July temperature in the Netherlands Pleniglacial, based on plant species. After Kolstrup (1979).
F. July temperatures indicated by Coleopteran faunas during the Devensian in lowland southern and central British Isles. Interstadials: UW, Upton Warren Interstadial Complex; C, Chelford Interstadial. After Coope (1977).
G. Tree pollen (AP) / non-tree pollen (NAP) curve for the Weichselian sequence at Grande Pile, Vosges. Interstadials: SG 1, St Germain I; SG 2, St Germain II. After Woillard (1975) and Beaulieu & Reille (1992).

The continuous climatic variation shown by the oxygen istope curve emphasises the difficulty of splitting the Quaternary into climatic units. Change continues during the parts interpreted as cold stages, with terrestrial expression of change as glacial advance (such as the *c*. 18–20 ka advance in north-west Europe, in Oxygen Isotope Stage 2, Figure 13.1A), and periods when ice sheets were apparently not so extensive. The changes are reflected in the presence of interstadial conditions in periglacial areas, non-glacial sediment sequences in glaciated areas, and permafrost development at times in periglacial areas. The variations within a cold stage are clearly exemplified by Donner's (1995) description of the geology of Scandinavia during the Weichselian (last) Cold Stage.

As well as variation of climate in time within cold stages, there is clearly also much variation in space, dependent on regional climatic conditions, as today. For example, the Early Weichselian interstadials in the Vosges are represented by forest with thermophilous trees (Figure 13.1G), whereas further north and west the forests are mainly coniferous, and further north still show birch and higher frequencies of non-tree pollen (Donner 1995; Mangerud 1991; Woillard 1978).

Significant regional factors also affect climate and vegetation. For example, Burn (1997) has pointed out that cooling of Holocene summer climates in the arctic coastal plains of Alaska and north-west Canada may result in part from coastal retreat and the southward movement of the influence of the Beaufort Sea. In the same region Porsild & Cody (1980) noted that 'Owing to the ameliorating influence of the great north flowing Mackenzie, the flora of its delta is surprisingly rich in southern species', with southern species mingling with truly arctic or maritime species. The July temperature isotherm of the area shows clearly the warmth of the Mackenzie Valley compared with the hinterland. A similar effect of the influence of the Rhine during the Weichselian of the Netherlands was suggested by Florschütz (1958) to explain the presence of southern species in the flora. Such migration of plants along rivers has been illustrated by Walas' (1938) description of the dispersal of plants down rivers in the Tatra forelands.

Such effects as these show how complex cold stage vegetation and floras may be, with regional climatic factors imposed on more local conditions affecting growth and dispersal. Add to this the variability of many of the taxa identified, and we see the difficulty of adducing cold stage climates and climatic change from the fossil floras. There appear fewer such problems when considering forest history derived from regional pollen rain in the Flandrian (Holocene).

Geological evidence

Before considering the biological evidence for cold stage climates, the main purpose of this chapter, a brief account of geological evidence for climate gives a necessary background for the biological evidence. The geological conditions and processes of the periglacial areas are the basis for the interpretation of climate. Those of Great Britain have been described by Ballantyne & Harris (1994), and in Ireland by Mitchell & Ryan (1997).

Sediments

Fluviatile sediments of main rivers are the major source of the fossil floras. These sediments are mainly gravels, with structures associated with present-day arctic nival or proglacial regimes of braided river systems (Bryant 1983a, b), indicating braided floodplains with high stages at the onset of the spring melt and possibly later. The similarities with the structures of present day arctic regimes indicates a similar hydrological cycle, with spring melt, but the lower latitude with increased summer length must have resulted in modifications. High stream activity following spring thaw is indicated by the relative coarseness of many gravel sequences, but considering the length of time involved in cold stages and the general shallow sequences found (say 2–10 m), aggradation must have proceeded in an intermittent fashion, interrupted by down-cutting as loads, base levels and hydrology changed.

The extensive deposition of loess in cold stages in the south and east England, though not in such depth as in more continental areas of Europe, indicates generally dry climates. As well as Devensian loesses, loess also is recorded immediately before the temperate Ipswichian Stage at Bobbitshole, Ipswich. Cover sand, coarser than loess, is more local than loess, but its presence indicates that dry sand sources were subject to wind erosion and redistribution. Aridity is indicated by the sand–loess primary filling of large contraction cracks (ice wedges) in Devensian gravel at Somersham, Cambridgeshire (West 1993), where a further sediment which has climatic significance is the calcite deposited in desiccation cracks in the clay–silt infill of troughs above the large thermal contraction cracks, indicating dry and arid conditions.

Solifluction diamictons in periglacial areas indicate local mass down-slope movement of water-charged surface sediments, a result of spring or summer thaw, in contrast to aridity suggested by the widespread aeolian sediments.

Figure 13.2. Thermal contraction cracks (ice wedge casts), indicating the presence of permafrost at two times within the aggradation of these Devensian gravels at Block Fen, Cambridgeshire. Photo A.J. Stuart.

Structures

The periglacial structures most significant for climate interpretation are those caused by freezing of the ground, producing permafrost. The freezing results in contraction, forming a polygonal pattern of cracks, which if they further expand to wedge-like features, are filled by sediments of local or regional origin, as described by French (1996) (Figure 13.2). In the Arctic, the development of permafrost is related to mean annual air temperatures of −6 to −8 °C, with much lower winter temperatures. But crack and wedge development are dependent on such factors as lithology of the penetrated sediments, snow cover, seasonal change of climate and time, and the interpretation of the structures is difficult, particularly remembering that we are dealing with their occurrence in the Quaternary at much lower latitudes that those occurring today at high latitudes. However, excluding seasonal cracks resulting from low winter temperatures, the presence of massive wedges indicates past permafrost at our latitude, with clear indication of very low winter temperatures.

Surface spring or summer thaw in areas with permafrost results in the formation of the active layer. The depth of the active layer, which may be indicated by the depth of freeze/thaw structural deformation of the sediment (cryoturbation) reflects the intensity of warming and also the nature of the sediment, since heat transfer is more rapid in sand or gravel than in finer sediments. So warmer soils are found on sand and gravel plains than in clayey uplands, which must result in a local differentiation of floras. Attempts have been made to relate depth of active layer to summer warmth (e.g. Williams 1975), but the interpretations are fraught with difficulty.

Polygonal systems and contraction cracks (Figure 13.2) have been described from the Devensian across the British Isles, from Co. Derry in Ireland to East Anglia, and from the southern Midlands to Scotland, mapped in Great Britain by Ballantyne & Harris (1994). There are also many records from earlier cold stages, back to pre-Cromerian times. The timing of these permafrost events, which must have been significant for the flora, is more difficult to ascertain. In the Devensian, evidence for permafrost has been found in the Early, Middle and Late Devensian, with most records in the Late Devensian. At some sites, e.g. Stanton Harcourt (Seddon & Holyoak 1985), regional permafrost was sustained over a period of gravel aggradation in the Middle/Late Devensian. At Somersham (West 1993), a period of massive contraction crack development followed the aggradation of gravels of probable Middle Devensian age, and appear to be of Late Devensian age.

The clearest evidence of changing climates and permafrost conditions in the Last Cold Stage comes from the Netherlands, where there have been detailed studies of long sequences of Weichselian fluviatile and aeolian sediments (e.g. Kasse *et al.* 1995; Ran & van Huissteden 1990; Vandenberghe 1992a,b, 1993) (Figure 13.1B, C). Two major periods of permafrost are indicated, one late in the Early Pleniglacial, the second in the Late Pleniglacial at around 20 ka BP. But there is also evidence for other episodes in the Middle Pleniglacial, together with frequent levels with freeze/thaw structures, erosional levels with desert pavements and a period of thaw lake development associated with the degradation of permafrost (Kasse *et al.* 1995). In the period 43–35 ka BP these authors have suggested a mean annual temperature of *c.* −4.5 °C, with a short period of a rise of 2–3 °C. within this time, indicating permfrost degradation.

In addition to the temperature changes, changes in precipitation have been identified from geomorphology and river development, with a conclusion that the Middle Pleniglacial (62–26 ka BP) was wetter than the Early and Late Pleniglacial (Vandenberghe 1992a,b).

The climatic sequence of the periglacial areas in the Devensian and Weichselian is obviously complex and variable. Moreover, it is often difficult to relate particular fossil floras to the climatic episodes identified by their geology. The cold stage floras described here must have responded to these changes, but the difficulty of analysis of such effects in terms of the very varied flora and the environments will be apparent. Kolstrup (1995), in a summary of palaeoenvironments in the north European lowlands between 50 and 10 ka BP, suggested that there were rapid and climatic and environmental changes during the Weichselian, often too brief for the establishment of stable vegetational communities and soil developments. This would compound the difficulties of interpretation, but the scenario could fit well with the nature and survival of the cold stage flora already described.

Biological evidence

Physical constants affecting development of periglacial processes and structures are known, so that certain aspects of cold stage climate, discussed above, can be reconstructed with some confidence. There are no such biological constants. The properties of species are variable, expressed in their taxonomy and physiology, which makes for difficulties in using species to define past climates. The question is: to what extent can vegetation and its component species reveal past climates and climatic change in cold stages?

The use of biological evidence for reconstruction of past climate is usually based on the present distribution of species, this being taken to have a particular relationship to present climate. But distribution is also controlled by historical factors, the ecology of the biotype concerned, competition and environmental factors, so that the use of present distribution is more complex than might appear. As already discussed in Chapter 12, there are important environmental variations to be considered in comparing present northern vegetation and flora with cold stage vegetation and flora at lower latitudes.

Environmental conditions in northern latitudes, as they affect plants, have been summarised by Billings & Mooney (1968) and Fitter & Hay (1987). The characteristic features are low soil and air temperatures, short growing season, low nutrient availability, wind, and water stress. In cold stage conditions at lower latitudes, there must have been very different conditions, with a greater length and favourable nature of the growing season, more favourable temperatures and soil conditions seasonally, and higher nutrient availability. Variation in snowfall at the lower latitudes will perhaps also affect distributions, allowing northern plants to range further south in areas with less snow. Such differences in environment between present northern latitudes and cold stage lower latitude conditions hinder making a straight relation between climate and distribution of a species.

A clear example of this is pointed out by Iversen (1954) in a discussion of the climatic significance of water plants in the Weichselian late-glacial lakes of Denmark. These were eutrophic, unlike most arctic and subarctic lakes in Scandinavia, so that caution was necessary when using species of eutrophic water (e.g. *Myriophyllum spicatum, Potamogeton compressus*) as climatic indicators, whereas other species (e.g. *Potamogeton natans, Typha latifolia, Nuphar* sp., *Nymphaea* sp.) were not so demanding and therefore more reliable climate indicators.

The effects of changing climate on species and communities have been studied in detail by those working on the possible effect of global climatic change on distribution of plants and animals (see e.g. Bradshaw & McNeilly 1991; Davis *et al.* 1998). Bazzaz (1996) has described the complexities of the relation between plants and environments in changing conditions and discussed plant successsion in relation to global change. It is against such a background that we should consider the biological evidence for cold stage climates.

Interpretation of climate

Two approaches to reconstructing past climate are through identification of past vegetation formations and of plant species. Formations are related to major climatic regions, for example the zonobiomes of Walter (1985); they can best be identified by palynology, the generally wide dispersal of pollen, identifiable often only at a generic or family level, revealing a regional aspect of vegetation indicating the formation. Macroscopic remains can give specific identifications, and species lists can then be used for more detailed interpretation of climate, based on the known behaviour (ecology, distribution) of the species at the present time.

Climatic change on the regional scale can more surely be identified by palynology, with changes of vegetation formation. Thus interstadials with forest within cold stages stand out clearly within the cold stage record. Lesser degrees of climatic change may be more difficult to identify, since the constitution of local plant communities can be affected by many factors both within the community through competition and through local environmental change, as well as changes in, say, regional seasonal precipitation or temperature. A combination of palynology and species lists then offers the best basis for interpretation of past climates, remembering the difficulties discussed above.

Estimations of past cold stage climate in north-west Europe

Estimations of past climates have been based on a range of palaeobotanical evidence, from identification of vegetation formations to the use of species identifications for more detailed reconstructions.

Comparison with the major zonobiomes of Walter (1985) via palynology may give a broad indication of climate, in terms of humidity (oceanic to continental) and temperature (temperate to arctic). Temperate deciduous forest can be distinguished from boreal forest, giving information on climate (if the vegetation is in equilibrium with climate). But there is a more ambiguous indication where herbaceous (grassland) vegetation is concerned, since it occurs in both arid and cold climates, giving rise to the steppe-tundra question discussed in Chapter 12.

In north-west Europe, identification of polar desert, steppe, tundra and shrub tundra in the Weichselian (Last) Cold Stage has been based on pollen and macro analyses (e.g. Figure 13.1D), and the identification then transferred to the lower latitudes in terms of July temperatures thought to control the present distribution of these formations (sometimes described as ecostratigraphical units) (e.g. Ran 1990). The problems of making such a straight transference of climates has already been discussed. Ran & van Huissteden (1990) have utilised the biological data, sedimentological data and periglacial phenomena to produce estimated temperature and precipitation curves for the Weichselian Middle Pleniglacial of the Netherlands.

More detailed indications of climate have been based on particular species. Schouw (1822), in his book on plant geography, used the term 'temperatursphaere' to describe the relation between a plant and its distribution in terms of temperature. Iversen (1944) refined the subject in his classic work on the relation of the distribution of the temperate evergreens *Viscum album*, *Ilex aquifolium* and *Hedera helix* in Denmark, calculating the 'thermosphere' (climatic envelope) of January and July temperatures which appeared to control their present distribution, and then using this data to calculate Holocene January and July temperatures according to the representation of the three species in the fossil record in the Holocene in Denmark.

Iversen (1954) also gave a remarkable and detailed account of the Weichselian late-glacial flora of Denmark and its relation to temperature and precipitation, based on the present distribution of the species. July temperatures were suggested, and conditions related to subarctic or warmer climates. Precipitation estimates were based on the presence of chionophobous species and on the presence of a steppe element thought to indicate rather dry summers.

Andersen (1961) described the Danish Early Weichselian sequence in terms of interstadials (*Betula, Pinus*) alternating with periods of severer climate. Pollen diagrams indicated vegetation types, which were related to July temperatures, via the present distribution of the species or genera concerned.

A climate curve of average July temperatures and vegetation for the Weichselian of the Netherlands was drawn up by Zagwijn (1975), based on palynology and the periglacial structures. Early Weichselian Interstadials were represented by forest pollen assemblages, with periods with higher non-tree pollen percentages taken to indicate subarctic parkland, shrub tundra, steppe-tundra, tundra and polar desert. Kolstrup (1979, 1980) estimated July temperatures for the Middle and Late Weichselian in The Netherlands from the presence of particular genera or species, based on their present distribution in Europe and in relation to July isotherms (Figure 13.1E).

The most detailed study of cold stage floras in Britain, by Bell (1969, 1970), was based mainly on macrosopic remains. Bell discussed the temperature requirements of the flora in much more general terms, with a discussion of the significance of the southern and steppe elements, but making no close estimations of climate, and referring to the difficulties of interpretation discussed above.

More recently, transfer functions have been applied to fossil pollen assemblages to estimate past climates. Modern pollen rain is related to contemporary climatic conditions to obtain a function which can then be applied to fossil assemblages to estimate the climatic character of the time (Bradley 1985; Huntley 1990).

Guiot *et al.* (1993) have applied such methods to estimate mean annual and July temperatures and annual actual evapotranspiration during the Weichselian Cold Stage, using the long pollen records at a number of western European sites. A particular problem encountered was the lack of good analogues for the cold stage spectra and the distinction between tundra and steppe vegetation, a subject already discussed. Since this approach is based on pollen taxa, with identifications mostly at a family or generic level, the contribution from a close knowledge of the species concerned in the vegetation is lost.

The problem remains of using present relationships of vegetation and flora to climate as a guide to past climate, the difficulties of which are illuminated by Davis *et al.* (1998) in their comments on the use of climatic envelopes in predicting the results of global climatic change on biota. In studies of more recent forest history, where analogues are more likely to be available in present vegetation, the transfer function approach is more likely to result in a realistic estimate, but for older herbaceous communities with an analogue problem, the information from pollen analysis needs to be supported by macrofossil studies.

Indications of climate based on vegetation

Stadial vegetation

As discussed in Chapter 12, the fossil record of the stadials indicates the presence of a regional herbaceous vegetation, with shrubs confined to local habitats such as floodplains. The absence of trees, some times even when the presence of thermophilous Coleoptera indicates warmer summers, has led to much discussion of why this should be so (e.g. Kolstrup 1990).

At present lack of summer warmth over the short growing season limits the growth of trees at the northern tree limit, with winter desiccation an additional factor. Winter cold is not necessarily a control, as witnessed by the taiga forests in areas with very low winter temperatures. Other factors which have been suggested to contribute to the absence of trees from tundra are the shallowness of the active layer, low soil temperatures and wind (Chapin & Shaver 1985). Wind has been considered an important factor in limiting tree growth in the tundra-steppe regions of Beringia (Yurtsev 1985).

At the lower latitudes of our area, with a potentially longer growing season and where conditions for tree growth might be expected to be more favourable, explanations based on the nature of the northern tree limit are not necessarily directly applicable. The fossil record (both flora and fauna) suggests wide expanses of grassland, and this immediately suggests climates where annual or seasonal drought is significant in determining the distribution of plants. An analogy may be drawn with the present steppe of central Europe and eastwards, where there are generally two annual drought periods, vividly described by Haviland (1926): one the physical drought of the later summer (after a rainy spring, or the snow melt of spring), the second the physiological drought of winter, associated with cold and wind. Steppe is characterised by a semi-arid climate, with a cold winter and a drought period in late summer. More than half the precipitation is in the summer, but soil water is insufficent to support forest. The relict steppe of central Europe is a result of low field capacity of the soils and high potential evaporation on south-facing slopes, with the dry periods shorter and more frequent than those of the steppe proper.

Similar conditions, characters of a continental climate, seem most likely to have played a significant part in limiting regional tree growth in the cold stage stadials. However, it is dangerous to draw too close an analogy with steppe (as with the present Arctic). The support of knowledge of soil development would be needed for a closer comparison.

Woodward (1987) points out that leaf mass, measured as leaf area index, will be under the control of the local hydrological budget, with a low leaf area index being expected in a dry environment. Grassland has a low leaf area index (<2), supporting the indications of a summer with a degree of drought.

The predominance of hemicryptophytes and cryptophytes in the life form analysis (Table 9.1) distinguishes the cold stage biological spectrum from present Arctic spectra, and is more comparable to spectra from temperate regions, as discussed in Chapter 9. This again points to the difficulty of comparing cold stage conditions to those of the Arctic in terms of climate, vegetation and flora. However, parallels have been drawn with arctic and subarctic vegetation formations (e.g. arctic tundra, subarctic park landscape, subarctic forest) to indicate July temperature changes in cold stages, as in early studies of the Weichselian in north-west Europe (Andersen, de Vries & Zagwijn 1960).

Interstadial vegetation

The distinction between stadial and interstadial vegetation has already been discussed (Chapter 12). In the Early Devensian, the Chelford and Brimpton Interstadials show forest interrupting the sequence of stadial herbaceous vegetation. This type of change in north-west Europe has been related to changes in July temperatures. July temperature estimations for Early Weichselian forest interstadials (and stadials) have been made by Zagwijn (1961) for the Netherlands and Andersen (1961), the values for forest conditions being > c. 10 °C. Andersen also discussed the problem of humidity. He suggested on the basis of an analysis of hydrosere changes that decrease in summer temperatures was accompanied by a rise in water level.

The forest assemblage of the Chelford Interstadial (*Pinus-Betula-Picea*) (Simpson & West 1958) was compared to that of the Fennoscandian conifer-birch region, with July temperatures 12 to 16 °C and winter temperatures −15 to −10 °C, mean annual temperature 2 to −3 °C, and an annual rainfall of 400–700 mm.

Andersen (1961) pointed out that rainfall in the interstadials must have been sufficient to support forest. The change from stadial herbaceous vegetation to interstadial forest vegetation may not only be an indication of July temperatures, but also a question of precipitation changes. If the stadial vegetation indicates a continental climate with summer drought, then it may be useful to compare this situation with the continental forest-steppe ecotone of eastern Europe, where soil-water conditions and summer drought control the major vegetation types, with summer drought promoting steppe (Walter 1985). In that case, the development of an interstadial may be the result of precipitation changes just as much as of changes in summer temperature.

Since no great differences have been found between the vegetation and flora of the Upton Warren Interstadial Complex and of the stadials, the question of the climate of this interstadial is not considered here but later, taking into account the faunal evidence of interstadial climate.

Indications of climate based on species

An identified species in a fossil flora normally means that that species was present in that locality at a particular time, regardless of any taphonomic bias to the assemblage. For interpretation of climate this is an advantage over the use of pollen spectra, particularly if the present distribution of the species and its relation to climate is known in some degree.

In considering the significance of species, it is useful to start with Iversen's (1954) comments in relation to his interpretation of Danish Weichselian late-glacial temperatures. He divided the flora into four groups: trees and bushes, water plants, shade-tolerant herbs and small bushes, and heliophilous herbs and small bushes. He regarded the first group as able to give reliable information about the macroclimate, the other groups being more strongly influenced by local conditions and microclimates. Water plants were also considered useful (provided edaphic conditions were known; see above), since water body temperatures may relate to regional climates, unlike more localised microclimates of terrestrial conditions. The last of Iversen's four groups, including light-demanding herbs of wide distribution today, was regarded as the most unreliable.

In terms of the cold stage flora, the poor representation of macroscopic remains of trees rules out their use in intepetation of stadial climates, though the interstadial presence of trees of boreal distribution is of course significant for climatic change. A large proportion of identified species are heliophytes, with a wide ecological amplitude, determined by genetic variability, phenotypic plasticity and varied regenerative strategies. Nevertheless, certain species have been used as temperature indicators.

A number of other variables affecting the use of present distributions of species to estimate past climates need to be considered.

Variation in habitats

Habitats favourable for species of more southern distribution exist in tundra today. Chernov (1985) points out that southern species of plants and animals penetrate north into the subarctic along well-warmed slopes, especially sandy ones (e.g. *Thymus serpyllum*), and that azonal habitat conditions (e.g. mires) allow the growth of southern species (e.g. *Caltha palustris*) in tundra. A related aspect of habitat width is in the variety present on floodplains, the drier soils offering suitable warmer conditions for southern species than occur on wetter plateau sites.

Aspects of control of present distribution

The present distribution of arctic plants is determined not only by their positive ability to survive the conditions imposed by low temperatures (e.g. short growth season, cold wet soils, low nutrient availability) but by such factors as competition and shade to the south. In suitable habitats, some species considered northern can be found in temperate areas (e.g. *Dryas octopetala*). Northern limits of temperate species may be further north in continental areas compared with more oceanic areas (e.g. *Trapa natans, Stratiotes aloides*), or conversely, more oceanic species extend further north to the west of the continent (e.g. *Potentilla anserina, Succisa pratensis*). Northern limits of plants clearly may vary with continentality of climate, not immediately discernible from fossil assemblages.

Analyses of distribution as a guide to climate

Chapter 11 described various analyses of the present distribution of species of the cold stage flora. Such treatment of the flora can only give a very general guide to climatic change. Data from Preston & Hill's (1997) analysis indicated an increase in the northern, wide-temperate, Eurosiberian and Eurasian categories, and a reduction in the temperate, southern, oceanic and suboceanic categories. The Ecological Flora Database analysis shows a similar result, with a reduction of the westerly and southerly species. The conclusion is that cold stage climates reduced the oceanic and southern elements of the flora and increased the more continental and and wide-spread elements, giving a continental aspect to the flora and climate.

Use of species

The present distribution of species in relation to climate has been widely used as a guide to cold stage climates, particularly to minimum July temperatures. Winter temperatures and humidity are more problematical, as Andersen (1961) pointed out. The latter can better be estimated from vegetation changes (presence of forest, heath) and water level changes indicated by sediment sequences. Iversen (1954) and Andersen (1961) have both provided detailed accounts of the problems of interpreting past climates from the present distribution of species, mainly in Europe. For a wider view of distribution the maps provided by Hultén (1958, 1962, 1971) and Porsild & Cody (1980) are of particular use in respect of circumboreal distributions.

For tree species, altitudinal and northern limits have long been used to indicate past temperatures, such as the use of the 10 °C July isotherm as controller of the northern limit of forest, then using this figure to distinguish July temperatures of stadial from those of interstadial times.

Iversen's (1944) estimations of temperatures in the Holocene of Denmark, based on the present and past distribution of pollen of *Viscum*, *Ilex* and *Hedera* have already been mentioned. He later (1954) extended climate interpretations based on species to the Danish Weichselian late-glacial, estimating summer and winter temperatures and precipitation. The approach was to take the present distributions, and relate them to present winter and summer temperature gradients. Andersen (1961) applied the methods of Iversen, estimating average July temperatures, using the present distribution in central and northern Europe of species found fossil. He considered that changes in herb pollen frequencies were an important guide to past temperature changes. Kolstrup (1979, 1980) applied similar methods, listing a range of indicator plants (pollen and macroscopic identifications) for estimating July temperatures, using also using ecological information on the species from Ellenberg's (1974) study of central European species.

Such approaches as these may be valid in more recent times in the Holocene, where the vegetation and flora can be more nearly related to the fossil assemblages. But when we are dealing further back in time with fossil assemblages with high non-tree pollen frequencies and non-analogue combinations, the question is whether we can use the Linnean species as a simple entity for transferring present climatic requirements back into the past. As we have seen, many of the species identified are very variable (Chapter 9).

The temperature control of the distribution of indicator plants has usually been based explicitly on their distribution in northern and central Europe. But many of the plants concerned have a much wider distribution. For example, the great majority of species listed by Kolstrup (1979) have a distribution outside northern and central Europe, in the Eurosiberian, Eurasian and Circumboreal categories of Preston & Hill (1997). Moreover, the taxonomic complexity of many cold stage species is an added complication in the interpretation. For example, *Armeria maritima* s.l. has been used as an indicator species for January temperature (range −8 to 8 °C) (Iversen 1954), but Bell (1970) points out the inconsistency of this with other indications of winter temperature in stadial times. In fact, the circumboreal *A. maritima* s.l. is distributed far to the north (Hultén 1958). What is needed is the association of the cold stage taxon/a to a present biotype(s) of the species in order to use it as a temperature indicator.

As an additional factor, it is probable that migration of biotypes in herbaceous communities in the open cold stage environments may have been very rapid, as witnessed by the recent rapid spread of plants along roadways (Coombe 1994), so that present non-European biotypes may well have been present in cold stages, as well as the non-British species identified.

The mixture of floristic elements

This mixture really underlines the difficulties of using present distributions of herbaceous species to interpret past climates. Stadial floras contain a variety of floristic elements (Table 13.1). Species of northern distribution have often been used to characterise floras as of arctic type, but the reality is that usually these floras contain more temperate species than northern species. For example, the occurrence of *Betula nana*, often described as an arctic-alpine (though this species reaches south to the Alps in Europe, with localities in lowland Germany, and to 55° N in western Siberia (Hultén 1958)), has been used to indicate the northern nature of the flora, even though accompanied by many temperate species.

A number of northern species nowadays occurs in more southern areas, where soils and lack of shade combine to give appropriate habitats, e.g. *Dryas octopetala* near sea level on limestone in the Burren in western Ireland and in Sutherland, *Saxifraga oppositifolia* near sea level in the Scottish Isle of Skye, and *Betula nana* in lowland bogs in northern temperate Europe. Iversen (1954), in describing the Danish Weichselian late-

Table 13.1. *Numbers of species identified by macroscopic remains at particular sites in the floristic categories of Preston & Hill (1997) (excludes non-British species); see Table 12.1 for category description*

Site, sample, radiocarbon date (ka)	Major biome category									Eastern limit category					
	1	2	3	4	5	6	7	8	9	1	2	3	4	5	6
Late Devensian															
Beetley BB1 16500	3	2	–	3	5	2	3	3	–	–	–	5	1	5	10
Somersham SY 18310	–	1	–	–	3	2	7	2	1	–	–	5	2	3	6
Somersham SAG 18750	–	2	–	1	1	–	1	–	1	–	–	2	–	–	4
Barnwell 19500	11	4	2	4	16	1	8	4	2	1	2	13	9	4	26
Late/Middle Devensian															
Block Fen 14	–	2	–	1	10	5	8	3	–	–	–	4	5	8	12
Block Fen 10	–	1	–	3	9	3	5	1	1	1	–	6	–	6	10
Middle Devensian															
Somersham SAD 2A 28020	–	4	–	2	5	3	3	1	1	–	–	3	3	6	7
Wretton WU B 20 29120	–	1	2	4	17	7	6	3	1	1	–	7	8	6	19
Sandy SD010301 34055	3	2	–	2	8	1	4	–	1	–	–	8	3	4	6
Upton Warren Interstadial															
Earith E7 42140	4	6	3	2	12	4	12	2	1	1	–	13	6	5	22
Upton Warren Band 2 41900	–	–	1	1	3	2	1	1	–	–	–	1	–	2	6
Upton Warren Band 3	3	4	3	4	18	4	7	2	–	–	2	11	5	6	21
Middle/Early Devensian															
Earith E9 >45000	4	3	1	–	10	3	4	–	1	–	–	8	5	4	9
Sidgwick Av. 40–50	–	2	2	3	11	5	5	3	1	–	–	5	4	5	18
Wolstonian															
Broome B	5	4	–	3	8	–	1	–	–	–	–	3	2	–	16
Cromer Complex cold stage															
Ardleigh J71–85A	1	2	–	1	7	1	5	–	–	–	–	–	4	4	9

glacial soils, refers to pioneer vegetation at the foot of mountain screes in the Villacher Alps of Carinthia, with a mixture of northern and southern species: *Dryas octopetala, Fraxinus ornus* and *Pinus nigra*. He also comments on the combination of arctic-alpine calcicolous heath and south-east European steppe species presently found in the *Helianthemum-Artemisia* heath of the limestone pavement of the Great Alvar of the Swedish island of Öland (Pettersson 1965), thought to have survived from late-glacial times.

The southern limit of northern plants is more likely to be controlled by habitat and competition than by climatic conditions. The same is not true for temperate species. Northern limits appear to be controlled by climatic conditions (see e.g. Edlund 1987; Woodward 1987; Young 1971).

The ocurrence together of northern and temperate species in cold stage floras may not then seem so surprising, since absence of competition and shade, with suitable soils, would favour the expansion of present-day northern species.

The gradients of drought and drainage in cold stage landscapes would greatly improve the variety of habitats available to plants preferring open conditions. Two contrasting conditions can be envisaged: one dry and continental on uplands and plateaus, the other wetter and warmer on the gravels of the floodplains, giving rise to habitats of great variety which would facilitate the floristic mix. The comparative rarity of the most northern species, compared with more temperate species, might result from the patchiness of suitable habitats. This leads to the perceived difficulties of climatic intepretation, with plants of different present geographical categories found in the same areas.

Interpretation has to be based on the present distribution of species found fossil, with inferences for climatic control of the present distribution. But habitats and climates were very different in cold stages from the present conditions. If we lived in a stadial, our maps of plant distribution would look very different from those prepared today, and we would have a similar problem of explaining temperate stage distributions; the explanation would involve development of forests and shading and the importance of competition, the reverse of what we have considered above.

Comments on species

Recognising these problems of the precise indications of climate by the species identified, such as the problem illustrated above by reference to

Armeria maritima, examples may be given of some species which give more general indications of climate. Further species relevant to the problem of interpretation of past climate are considered in Chapter 12, while Iversen (1954, 1973), Andersen (1961) and Kolstrup (1979, 1980) discuss the significance of a wide range of species.

Typha latifolia, easily recognised from the pollen tetrads, is considered a thermophile, requiring a minimum July temperature of 12–15 °C. (Iversen 1954; Kolstrup 1980). In the cold stage record the pollen has been recorded sparsely in the Early and Middle Devensian stadials, in the Early Devensian interstadials, in the Early Wolstonian and in older stadials. Kolstrup (1979) recorded *T. latifolia* in the Netherlands Weichselian Pleniglacial, the Hengelo Interstadial and Moershoofd Interstadial Complex. The circum-boreal distribution (map, Hultén 1962) indicates the warm summers envisaged by Iversen, but cold winters (January, < -10 °C) can certainly be endured, the plant having a wide continental distribution. A continental climate with warm summers would evidently satisfy the requirements for the presence of the plant in stadials.

Helianthemum canum. This very variable species, in Preston & Hill's (1997) Mediterranean-montane category and recorded in many Devensian stadial sites, has a remarkable occurrence in the Late Devensian site at Somersham, Cambridgeshire (West *et al.* 1999), where macroscopic remains were present at 18,750 years BP, south of the margin of the Late Devensian ice margin. The present species (ecology described by Griffiths & Proctor (1956)) demands summer warmth, and is found from near sea level in the Burren in Ireland to 1800 m in South Tirol, indicating a wide range of habitat accompanying the variability. The Late Devensian records suggest warm summers at least, and an ability for the biotype concerned to withstand cold winters.

Lonicera xylosteum. Pollen of this doubtfully native species was recorded in the Devensian Chelford Interstadial, in spectra indicating coniferous forest (Simpson & West 1958). A continental climate is indicated.

Potamogeton spp. and *Groenlandia*. As water plants, these may be expected to give good indications of thermal climate, as discussed by Iversen (1954) and Szafer (1954). Table 13.2 summarises the Preston & Hill categories for recorded *Potamogeton* spp., and shows the variety of distribution types

Table 13.2. *Potamogeton spp. and Groenlandia: species recorded in five or more macroscopic samples and their floristic categories (Preston & Hill 1997) (See table 12.1 for category description)*

Species	No. of samples	Major biome category	Eastern limit category
Potamogeton natans	28	5	6
P. gramineus	18	5	6
P. alpinus	28	4	6
P. praelongus	24	4	6
P. pusillus	9	8	6
P. obtusifolius	7	5	6
P. berchtoldii	13	5	6
P. acutifolius	5	7	3
P. crispus	30	8	5
P. filiformis	80	4	6
P. vaginatus (non-British)	13		circumboreal
P. pectinatus	28	6	6
Groenlandia densa	53	7	3

represented by fossils. The large number of Eurasian and Circumboreal species and the spread of species in the Major Biome (northern to southern) categories are striking. It is impossible to pick out, say, a southern species (e.g. *P. crispus*, *Groenlandia densa*) or a northern species (e.g. *P. filiformis*) and state that its presence indicates a particular thermal climate. Rather an assemblage of species has to be considered as a whole for the purpose of climate interpretation.

Bell's (1969) southern and steppe elements, listed in Table 11.2, reinforce the idea of the continental nature of the climate.

From these considerations it would seem difficult to give indications of stadial January or July temperatures with any degree of precision from the presence of a particular species. Assemblages may offer the best hope, but clearly much more information is needed concerning the ecological width and subspecific taxonomy of species, with corresponding knowledge of the fossils. The continentality of the cold stage climate, together with the rarity of recorded oceanic or suboceanic species, portrays an important character of the climate, but the degree of summer warmth or winter cold is much more difficult to assess. Against this background, the assessment of minor climatic changes within stadia appears difficult.

Supposed discrepancies in climatic interpretation of contemporaneous floras and faunas

This very interesting subject was raised by early interpretations of Weichselian late-glacial climates in southern Scandinavia (see Iversen 1954). Cold conditions were suggested by the terrestrial plants, but more favourable conditions by the hydrophytes and freshwater molluscs. Wesenberg-Lund (1909) supposed that this contrast resulted from the higher angle of the sun at the lower latitudes resulted in higher temperatures of the shallow late-glacial lakes compared with those of the Arctic, with the consequence that aquatic flora and fauna would be unsuitable as indicators of macroclimate. Iversen concluded that the present distribution of water plants did not support Wesenberg-Lund's theory, since latitude was not a clearly controlling factor, and that therefore water plants could be valuable as indicators of thermal climate. However, the apparent contrast of the evidence from terrestrial plants and hydrophytes remained. This may well be the result of the more rapid migration of hydrophytes as climate ameliorated, compared with the migration of the trees which eventually suppressed the herbaceous vegetation by competition, as Iversen suggested.

A similar problem has arisen in the discrepancy between the evidence for climate from the flora and fauna of the Devensian Upton Warren Interstadial (Figure 13.1F). The flora and vegetation remains the herbaceous vegetation characteristic of stadials, with the assignation to interstadial status based on a coleopteran fauna described as temperate compared with cold-adapted faunas below and above. Such faunas have been described from a number of Devensian floodplain gravel sites, especially those of the Thames Valley at Isleworth and South Kensington (Coope & Angus 1975; Coope *et al.* 1997). At these sites the climatic significance of the beetle assemblages was determined by reference to their present distribution, using the Mutual Climatic Range method (Atkinson *et al.* 1987) to infer mean temperatures of warmest month 16–17 °C and coldest month -4 ± 6 °C or -3.5 ± 2.5 °C. These temperatures resemble those of southern England today, except that the winter temperature is somewhat lower.

Coope (*et al.* 1997) discussed in detail the significance of the contrast of evidence. He suggested that since the interstadial climate indicated by the beetles was 'thoroughly temperate and moist' the explanation should be sought in a sudden climatic change to which the beetles could respond rapidly, but to which there was a lag in response by trees because of their slow migration capability. Coope further comments that the delay may have been encouraged by a grazing factor (large mammals frequently recorded at sites) and by soil conditions being poorly developed.

Other causes considered by Coope are wind and climatic dryness leading to steppe conditions. Regarding the former, there is no evidence for strong aeolian action at the time. Regarding the latter, Coope comments (South Kensington site) 'that all sedimentary and palaeontological evidence points to a river that flowed throughout the year, which in a temperate environment was presumably fed by fairly continuous precipitation'. Vandenberghe (1992b) also considered on the basis of sedimentology and geomorphology that precipitation during the Middle Pleniglacial in the Netherlands continued throughout the year.

This explanation has to be reconciled with the climatic interpretation of the presence of grassland, discussed above, where a continental climate and annual or seasonal drought was suggested as the cause of the absence of trees. If there had been a climatic change to more temperate and moist conditions, an expansion of *Betula* and *Salix*, pollen of which taxa is present in very low frequency, would be expected. The increase in the diversity of the herb flora and in the representation of Umbelliferae pollen and macroscopic remains reported at certain sites may be a result of local changes of plant communities. Many of the taxa concerned occur in many Devensian stadial assemblages; this problem of interpreting minor changes in herb pollen assemblages in climatic terms is discussed above.

In contrast to these differences, it is notable that the evidence for climate from flora and fauna, the Devensian late-glacial evidence from beetles and pollen tends to converge to give a similar picture of temperature change (Lowe & Walker 1997a).

An interesting discussion reflecting this problem has been raised by Sher (1997), who has contrasted a situation in the Siberian Arctic where in supposedly afforested ('interglacial') periods the vegetation has been described as having analogues with present taiga vegetation while the insect fauna is a non-analogue mix of tundra and steppe species. He concluded that the vegetation was in an open birch woodland, with grassland and a richer herb content than the present taiga, giving a non-analogue insect fauna.

A particular problem in such discussions as these is the climatic terminology. The use of the term temperate, which generally applies to the areas of the Earth between the polar regions and the subtropics, covers a great variety of climate, from oceanic conditions on the western side of continents to continental conditions in the hinterland of continents. The climatic diagrams of Walter (1985) illustrate this variety. As important as summer and winter temperature is the seasonal distribution of precipitation, which was discussed above in relation to the presence of grassland, and needs to be considered in any reconstruction of climate. The use of the term 'arctic' is also confusing, leading to the supposition of an arctic climate, which should not be directly imposed on our latitudes. As we have seen and tried to stress in the discussion of stadial flora and vegetation, the use of present distribution to identify past climate has its difficulties.

The subject of insect populations and environmental change has recently been discussed by Lawton (1995). Davis *et al.* (1998) have discussed the difficulty of use of climatic envelopes in relation to global warming and *Drosophila* populations. Andersen (1993, 1996) has also commented on the problems of the use of climatic envelopes of beetles for reconstructing climate, while Williamson (1996) has discussed the problem of climatic matching of plants and animals in more general terms. Climatic change has a complex effect on cold stage biota and it is not at present clear to what degree climatic envelopes should be used to reconstruct past climates.

The reconciliation of the flora and fauna evidence for climate suggested by Coope has, then, its difficulties. A possible alternative explanation may lie in seasonal changes of climate, such as increased summer temperatures which allow the spread of thermophiles, while continentality continues to exclude trees.

The question of interstadial climates

Climatic changes in north-west Europe may involve temperature and precipitation changes, both mean annual and seasonal. They may be expressed in terms of the present north–south gradient of climate (or vegetation associated with that climate, arctic to temperate), as with Jessen & Milthers (1928) definitions of interglacials and interstadials which relate to summer temperatures. Or they may be expressed as a west–east gradient of climate (oceanic to continental). The former is the gradient most commonly used for expressing climatic change during the last cold stage, with temperature and precipitation estimated from sedimentology, periglacial structures and palaeontology, e.g. the curves for mean, January and July temperatures and for humidity/precipitation given by Vandenberghe (1992a,b) during the Netherlands Weichselian Pleniglacial (Figure 13.1B). However, change seems more likely to take place along both gradients. It is the complex interplay resulting which will in part determine changes in the biota and sediments, with mean temperatures estimated from permafrost evidence the most reliable of the interpretations. The importance of the west–east gradient has perhaps not been fully realised, though Sher (1997) has shown its significance.

Early Devensian/Weichselian interstadials

These are clearly identifiable in the pollen record by the expansion of forest at many sites in north-west Europe (Behre 1989; Mangerud 1991). The two major interstadials, the Brørup and the Odderade, show a gradient from the Vosges (Figure 13.1G; Woillard 1978), through northern France (Emontspohl 1995) to Britain and northern Europe. To the south-east temperate tree genera are present with conifers; to the north and west boreal coniferous forest is present. Mean July temperature curves have been drawn, based on the presence of forest. However, a change in seasonal distribution of precipitation may also have assisted the spread of forest. The pollen curves of *Artemisia* and Gramineae in the Grande Pile pollen diagrams of Woillard (1978) show marked expansion in the stadials, suggesting a change to continentality in those periods, which may have been reversed in the interstadials without a considerable change in July temperatures.

Middle/Late Devensian/Weichselian interstadials (excluding the late-glacial)

Interstadials have been recognised in the the Middle Devensian and in the Weichselian Pleniglacial of the continent in the Netherlands and northern Europe. The classic sequence in the Netherlands (Figure 13.1D) included the Moershoofd (50–43 ka BP), Hengelo (39–37 ka BP) and Denekamp Interstadials (32–29 ka BP) (see Vandenberghe 1992b). These were associated with accumulation of organic sediments, but with increased knowledge of the sedimentology, palaeontology and radiocarbon ages, it appears that organic sedimentation occurred throughout the Pleniglacial, with a concentration around 42–40 ka BP associated with the changing palaeohydrology, controlled by the nature of the river system (van Huissteden 1990; Kasse *et al.* 1995). Kasse *et al.* (1995) in their study of the period 43–35 ka BP in the Netherlands found that there was the repeated occurrence of ice-wedge casts during the period and suggested a mean annual temperature of *c.* 4.5 °C, but with summer temperatures 2–3 °C higher between 38.7 and 36.9 ka BP, indicated by permafrost degradation and changes in the flora (based on the temperature indicator taxa of Kolstrup (1980). This is within the period of the Hengelo Interstadial. The vegetation in the area studied was interpreted generally as shrub tundra. Organic sediments with palaeobotany indicating shrub tundra also characterised the Oerel and Glinde Interstadials of Behre (1989) in north-west Germany (Figure 13.1D), placed in the early part of the Middle Pleniglacial.

The identification of climatic variation and of interstadials (and their definition) within the north-western European Pleniglacial clearly requires knowledge of sedimentology, palaeontology and frost structures, in the way developed by Kasse *et al.* (1995). Radiocarbon dating, also required, needs close attention, with problems at this age of sediments of contamination and reworking.

In Britain the definition of Middle/Late Devensian interstadials has the problem of the conflicting evidence of the biota, discussed above. The Upton Warren Interstadial Complex may correlate in part with the Hengelo Interstadial. In addition to correlation based on radiocarbon dates and palynology, as Coope *et al.* (1997) point out, the correlation of the Upton Warren period and its temperate insect fauna with the continental interstadials requires the investigation of their insect faunas.

14

A wider view of cold stage biota

Previous chapters have described the constitution and phytogeographical relationships of the cold stage flora, habitats, vegetation and climatic indications. Considering that the fauna may be largely dependent on and therefore closely related to the nature of the flora and vegetation, it is of interest to extend the view to include other components of the biota, making first some general points and then discussing particular groups.

The response of communities or species to temporal changes such as succession or climatic change has been viewed in two distinct ways. The Clementsian view, based on observations of plant succession, is of the community as a dynamic organic entity, with a history of its own. The Gleasonian view is of each species responding individually to change in accordance with its particular ecological requirements. Watt (1964) and Graham et al. (1996) have considered these viewpoints, the former in relation to experience of recent plant communities, the latter in relation to Late Quaternary vertebrate faunas in North America. Watt (1964) showed how both viewpoints contributed to the study of vegetation, with dominants in the plant community exerting an effect on the pattern of the community and its history. Grichuk & Grichuk (1960) showed how climatic change resulted in new combinations of plant formations in the Russian Plain around the time of maximum glaciation, a Gleasonian view.

In terms of the Quaternary history of biota, these viewpoints regard present communities either as having a discrete history or as being temporary aggregations of species under particular conditions of climate, environment and history. In terms of cold stage biota, we have seen that the flora is complex, with elements from different and distinct present-day formations, giving a mix of geographical elements. Such a mixing is not confined to plants, but is shown by vertebrate faunas, beetle faunas and mollusc faunas of the Last Cold Stage (Devensian, Weichselian, etc.), as discussed below. The species composition of such assemblages, with no present-day analogues, clearly expresses the importance of the individual in the Gleasonian sense.

The heterogeneity of cold stage habitats (Chapter 10) must have contributed to the breadth of the assemblages. This heterogeneity, together with the open landscapes, is likely to have aided rapid migration of species of all components of the biota, including herbs, vertebrates, insects, molluscs, etc. The rapid migration of species following forest clearances and changing agricultural practice in Europe in and after the Neolithic illustrates this aspect of cold stage biology.

With the spread of forest in our latitudes at the end of the Last Cold Stage, heterogeneity was reduced, and communities and environments emerged in the lowlands and mountains which are more comparable to those of today.

Vertebrate faunas

Stuart (1982) considered the typical vertebrate faunas of the cold stage stadials in Britain to be a mix of extinct taxa, taxa now confined to arctic or (rarely) alpine regions, taxa nowadays characteristic of steppe, and taxa now confined to regions further south. Such assemblages are found in the Anglian, Wolstonian and Devensian Stages. He comments that they reflect climatic and vegetational conditions not represented anywhere in the world today, and that though the faunas are less diverse than those of the temperate stages, the abundance of bones (especially *Bison*) in the Devensian indicates that at certain times the vertebrate biomass may have been greater than that of any time in a temperate stage. This argues for a high productivity of the cold stage vegetation (grassland) at these times.

The Devensian fauna has been described in detail by Stuart (1982, 1995). The faunas vary in time, with more abundant *Bison* in the Early and Middle Devensian (Upton Warren Interstadial Complex?), but with poorer faunas in the Late Devensian. The distinction is not clearly reflected in the flora.

A similar distinction within Devensian faunas has been made by Currant & Jacobi (1997). They distinguish a Middle Devensian fauna, the Coygan-type fauna, and consider it of OIS Stage 3 age. The fauna is rich in large herbivores, indicating a productive environment. It is considered to be of a type characteristic of the later Quaternary of central Asia north of the Himalayas, marking the westward extension of extreme continental conditions, with arid environments and extremes of seasonal temperature. An older Banwell-type fauna is more restricted and is considered to be of OIS Stage 4 age.

Stuart (1982) points out that in general the Devensian fauna is very similar to that of the last cold stage of the continent, and is consistent with the presence of a wide land-bridge and free immigration from the continent. The faunas from central Europe older than the late-glacial show a stronger steppe component with, for example, a greater number of records of the steppe antelope *Saiga* than the single Middle Devensian record from Britain. Broadly similar assemblages occur from France to Siberia, emphasising the wide sweep of herbaceous vegetation across Eurasia during the Last Cold Stage (the Mammoth Steppe of Guthrie 1990), and probably the previous cold stages.

The strongly continental type of climate indicated by the faunas tallies with the conclusions reached from a consideration of the fossil record of the plants. Since the herbivores depend on the vegetation, this is hardly surprising. This relationship between large herbivores and vegetation in the Weichselian late-glacial of Denmark was discussed by Iversen (in Degerbøl & Iversen 1945), in respect of finds of *Bison*, the presence of grassland and the record for steppe elements such as *Artemisia* and *Ephedra*, while Guthrie (1990) has treated the relationship on a wider scale in his discussion of the Mammoth Steppe.

An analysis of Late Quaternary mammal faunas in North America by Graham *et al.* (1996) illuminates the nature of cold stage assemblages. During full-glacial (Wisconsin) times (20,000 to 15,000 years ago), in addition to the presence of species now extinct, tundra and boreal forest species ranged further to the south, species of eastern deciduous forest extended westwards to the Great Plains, and grassland species extended eastwards. Some southern species were not displaced southward during the full-glacial. In the following late-glacial and late Holocene times new communities emerged as climate ameliorated, gradually assuming their present composition in the last few thousand years. These changes were accompanied by a reduction in the environmental heterogeneity of the full-glacial. From their analysis Graham *et al.* (1996) concluded that species of the mammal communities responded to Late Quaternary climatic change in a Gleasonian manner and the effects of climatic change have to be studied via individual species rather than by species associations.

Coleopteran faunas

The changing composition of beetle assemblages of the Devensian, described comprehensively by Coope (1975, 1977), indicate clearly a response, believed to be rapid, to regional climatic change. They indicate much detail of climatic change, and vary from restricted assemblages associated with Late Devensian colder conditions to those of the Upton Warren Interstadial Complex with more diversity and indications of climatic amelioration. These latter are the assemblages discussed in the previous chapter which show the flora and beetle discrepancies in climatic interpretation.

While the more restricted faunas, e.g. of the Late Devensian Barnwell site (Coope 1968), show a community containing low temperature stenotherms, other assemblages, such as those at Upton Warren (Coope *et al.* 1961) contain a mix of species of present-day northern, southern and continental species. The more restricted faunas of cold-tolerant species also sometimes contain species of more southern distribution. An interesting feature of assemblages of the colder parts of the Middle and Late Devensian is the presence of eastern Asian species, including a Tibetan species (Coope 1973), together with species now arctic-alpine. Range changes must have been very extensive, as with other components of the cold stage biota.

In many analyses of the flora and Coleoptera indications of climate are similar, e.g. in the Devensian Chelford Interstadial (Coope 1959) and in the Devensian late-glacial (Lowe & Walker 1997a). But discrepancies, as with the flora and beetles of the Upton Warren Interstadial Complex, have been reported from other sites, for example in sandy terrace sediments at Wretton (Early Devensian; Coope 1974) where pollen evidence for interstadial forest is accompanied by a restricted beetle assemblage indicating a barren environment. A Late Wisconsin site in river terrace deposits in Ontario, described by Ashworth (1977), presents a similar problem. Here pollen evidence indicates regional spruce forest with temperate trees, but the Coleoptera are a mix of faunistic elements with no known analogue with arctic-alpine, boreal, and eastern species. The macroscopic plant remains reflect the varied habitats of the valley, with again a mix of species with low arctic and boreal distributions. Ashworth (1977) suggests that the valley, in contrast to the regional forest, contained a variety of habitats with localised microclimates, as do certain areas today on the north and east shores of Lake Superior which contain a mix of species.

The question then becomes whether the mix of species is associated with a regional climate or with differing thermal environments within the catchment containing the fossil assemblages. Ashworth's (1977) explanation of the mix of species may also apply to the Wretton case described above, with terrace surface communities very different from regional forest communities. Studies are needed of the differentiation by habitat of localised communities and the fossil assemblages they produce, i.e. the taphonomy of the assemblages. In addition, the problems of relating distribution to climatic envelopes which are then used to postulate past climates, discussed in the previous chapter, need consideration.

Non-marine molluscan faunas

There is a substantial record of non-marine molluscan assemblages through the Devensian, with Early, Middle and Late Devensian faunas recorded. It is thought that faunas were able to respond rapidly to climatic change, not necessarily directly, but through other factors such as vegetation change (Preece 1997). Mollusca give valuable and clear indications of local habitats, such as degrees of shading, dampness or aquatic conditions, but the present wide distribution of many species found fossil makes an interpretation of thermal conditions difficult. The use of other methods of climatic interpretation has been discussed by Lowe & Walker (1997b). These include using transfer functions and present-day assemblages and the use of varying oxygen isotope ratios in shells.

The Devensian assemblages (Kerney 1977; Preece 1997) vary from restricted faunas of relatively few species in the Late Devensian (e.g. Somersham; Preece in West *et al.* 1999), to more diverse faunas of the Upton Warren Interstadial Complex (e.g. Ismaili Centre; Preece in Coope *et al.* 1997). In the Late Devensian assemblage at Somersham, no strictly arctic or alpine species were recorded in the mainly freshwater fauna. Species of the Middle Devensian Upton Warren Interstadial Complex faunas live in Britain at the present day and are common holarctic or palaearctic species, confirming the temperate nature of the climate indicated by the Coleoptera.

In the Early Devensian at Wretton (Sparks in West *et al.* 1974), the diverse fauna includes southern freshwater species, as well as cold-indicating species, such as *Columella columella*. Some of the southern species range east into Siberia, which led Sparks to suggest that cold winters and warm summers might explain the combination of species, a continental climate already discussed in relation to the flora. Sparks also supposed that the fauna had more affinity with the fauna of the preceding temperate stage than the more restricted later Devensian faunas.

It is notable that *Pupilla muscorum*, a holarctic species generally of dry grassland and sandy places, reaching 70° N in Scandinavia, is a common species of Early, Late and Middle Devensian faunas. It is a characteristic dominant of the cold stage European loess faunas, and its presence tallies with the evidence for grassland in the Devensian stadials.

A further relevant point has been made by Kerney *et al.* (1980), who describe mollusc faunas of the early late-glacial in south-east England which have a 'peculiar and diagnostic character' with a mixture of biogeographical elements that has no modern analogue. Kerney (1963) also notes that it would perhaps be a mistake to interpret late-glacial assemblages in the light of the present-day ecology of the species in temperate Europe, and that there is some evidence that climatic differences may alter the relation of a species to its environment. He points out, for an example, that in mountain areas certain molluscs which usually live in very diverse habitats may come together in close proximity in open country, perhaps partly because at high altitudes average temperatures remain low, preventing too rapid evaporation of ground moisture in open habitats, so permitting the existence of species which would normally shun such places. The observation underlines the problem of using present-day ecology of a species for interpretation of past climates.

Diversity of biota, habitat and climate

The presence of mixed ('non-analogue') faunas is one of the characteristic features of the cold stage faunas discussed above, as it is of the cold stage flora. This diversity is matched by the diversity of habitats available, a point stressed by Graham *et al.* (1996) in relation to the vertebrate fauna, and which is also evident in the nature and habitat requirements of the flora. The diversity of habitats, and the associated ecotypic diversity of the plants (e.g. Crawford 1997) is seen in present arctic (and alpine) areas and is related to variation of sediment and soil type, insolation, and drainage. It is apparent in the remarkable diversity of the flora of Wrangel Island in the Siberian Arctic, where the continental climate and the abundance of carbonate rocks support what are described as relics of the tundra-steppe communities of the last cold stage (Vartanyan *et al.* 1993). Transmitted to lower latitudes, with the absence of shading forests, diversity of microclimates may be even greater, contributing to the mixing of biota during cold stages.

The genetic resources available for coping with cold stage climates and climatic change are not readily apparent from specific identifications of the biota, or from the present-day distributions of the species, observed over perhaps a hundred years of ecological and mapping endeavour. The presence of the mixed biota cannot then be considered surprising. But mixed biota are not always a characteristic of cold stage biota. Restricted faunas of vertebrates, beetles and molluscs are present at various times, as described above. These are considered periods of severer climate, associated with particular times in the Devensian or with OIS stages 2 and 4, in contrast to the more diverse biota found in the Middle Devensian or correlated with OIS stage 3.

On the other hand, the stadial regional vegetation of grassland indicated by the pollen assemblages, with few specific identifications of Gramineae, does not indicate climatic change so clearly. This may be because grassland is a vegetation formation known from the Arctic to the tropics and not allied to a particular climate. The measurement of climatic change via different elements of the biota may help to explain the difference. The regional vegetation is a measure of regional climate, with 'above ground' climates controlling the nature of the plant life. Near and on the ground, microclimates and habitat diversity become important, as evident from Kerney's comments on molluscs mentioned above. These are the habitats which the fauna enjoys, and which are represented in the fossil assemblages, as with the plant macrofossil assemblages. To the aspect of cold stage conditions given by the regional vegetation must be added the diversity of local habitats and communities indicated by the animal and plant macrofossil assemblages.

15

Origin and fate of the cold stage stadial flora

Towards the end of the Tertiary in the northern hemisphere, a cooling of the climate led to the replacement of broad-leaved and coniferous forest by more open vegetation in the areas now occupied by tundra (Bliss 1981b; Chernov 1985; Wolfe 1985). Tundra may therefore be considered the youngest of present biomes, with the flora originating in the Late Tertiary and Early Quaternary from the open communities of Late Tertiary taiga forests, from alpine floras of the northern mountains, and perhaps also from the open communities of the steppes of the continental interiors.

With the advent in the Quaternary of periods of massive glacierisation from northern centres, forest limits retreated southwards, promoting a flora which may have contained species from unglaciated areas to the north, species from neighbouring mountain ranges and wide-ranging herbaceous species which took advantage of the open ground. On climatic amelioration and readvance of forests northwards, such floras must have become differentiated, an arctic component combining species of unglaciated arctic areas with species able to migrate north and survive, and an alpine component retreating to the mountains and a wide-ranging group surviving in the new temperate areas. Chernov (1985) considered that about 40 per cent of the resultant present arctic plant species were circumpolar in distribution, a reflection of the lack of barriers to migration in the open northern communities compared with further south.

These great climatic changes were repeated many times in the Quaternary. There is evidence in eastern Eurasia (Beringia) that in the later cold stage climates became more continental in character, with grassland biomes and their characteristic grazing fauna taking on the steppe-tundra aspect (Bliss 1981b; Giterman 1985). In north-west Europe, the vegetation history of the Early Quaternary indicates frequent climatic fluctuations with less well-developed cold episodes; a greater abundance of Ericales pollen is often present in both cooler and warmer times, suggesting a more oceanic climate (Menke 1975). After about 700,000 years ago the fluctuations became more obvious, with well-marked cold and temperate stages. It is to the earliest of one of these cold stages that the oldest British cold stage flora belongs, that of the Beestonian. By that time the characteristic cold stage floras of the Middle and Upper Pleistocene had evidently evolved, the previous million years of the Pleistocene, with apparently less intense climatic changes, providing the evolutionary background.

The changes of climate give a background of great complexity when considering the origin of present distributions of species and of cold stage floras, for example, on the arctic or alpine origin of species common to both. The origin of arctic and boreal distributions has been much discussed in the past (e.g. Hultén 1937, 1962). Matthews (1937) considered this problem in relation to the division of central European boreal and arctic-alpine elements into the 'historical northern' and 'historical Tertiary' of Kulczynski (1923). The former are regarded as migrating from the north, the latter are members of an original Tertiary montane flora. Hadač (1963) likewise discussed the origin of the arctic flora of Spitsbergen, distinguishing centres of origin, arctic and montane, for groups of species. These matters need consideration in the light of the improved knowledge of cold stage floras and the distributional history of the species in the Quaternary.

Chernov (1985) commenting on the nature of the present northern fauna and flora, pointed out

. . . that animals and plants filling differently formed intrazonal biotopes but of southern providence and not being specially adapted to living under arctic conditions are able to cross the climatic boundaries and settle in the tundra landscape. This is an ecological pathway of adaptation to the environment, which in the choice of suitable habitats also includes the avoidance of unfavourable influences.

This illustrates the way in which evolution of biotypes may occur in the past in the course of climatic change, and it also illustrates the co-existence of arctic and southern elements of the flora near the southern boundary of the arctic or subarctic. An element of active evolutionary change through selection of biotypes may then accompany migration of species.

This is the background against which we must consider the cold stage stadial flora. As we have seen, this flora is not an arctic flora and bears little similarity with the present arctic biome in terms of climatic conditions, fauna and flora. With the onset of stadial conditions and loss of forest, elements of an arctic flora move southwards and mix with elements of a temperate flora which could withstand the severer climatic conditions of the stadials, expressed as changes in seasonal precipitation and temperature conditions. The disturbed habitats of the stadial environment would favour both these components of the stadial flora, with the encouragement of the variability of many of the species (Chapter 9) and the selection of biotypes.

The changes in flora and vegetation from the last temperate stage (Ipswichian) to the Early Devensian were analysed by West *et al.* (1974) at the Fenland margin sites of Wretton and Histon Road, Cambridge, and the pollen and macro taxa occurring at both times were tabulated. Against a background of grassland at both times, many taxa appeared to have survived the changes of environment (Table 11.4).

In more general phytogeographic terms, and accepting a trend to continentality, the climatic change would lead to the following in our area: species of the arctic migrating south ('historical northern species'), alpine species moving north ('historical Tertiary species'), continental species moving west, and widespread species of sufficient ecological width staying put. Atlantic species would be restricted, only extending their area if oceanic conditions developed more widely, as perhaps in the forested interstadials.

Discussions of the history of the present British flora and phytogeography have in the past raised questions such as those of perglacial survival on nunataks or of total destruction (*tabula rasa*) of the flora. Many of these have now been resolved by palaeobotanical studies. Godwin (1975), in discussing these matters in detail, pointed out that it is not necessary to invoke perglacial survival to explain problems of distribution. The record of plant fossils through a cold stage such as the Devensian (Table 5.5), including assemblages from the time of the Late Devensian ice advance, makes it unlikely that there was ever a period when the flora was extinguished.

The extinction of the stadial flora as the post-glacial (Flandrian, Holocene) forests spread has been well described by Iversen (1954) and Godwin (1975). Temperatures improved, but a more critical change for the many light-demanding plants of the flora was loss of light by shading. Aquatic plants in their 'open' habitats suffered less, the cold stage aquatic flora surviving. The same may apply to marsh or fen species.

Later in the post-glacial, the development of upland mires further restricted the essential open habitats, as did the podsolisation of the fresh soils favouring many cold stage species. Species found refuge in areas where local environments, including soil conditions, prevented tree growth. Such areas included dunes, coasts, at altitudes in mountain areas, screes, cliffs, areas where peat growth never developed in the post-glacial, areas with shallow base-rich soils with open plant communities and bare rock pavements. The last occur in the British Isles in the Pennines and in the Burren in Co. Clare, where plants considered as arctic-alpine are found with species of continental distribution. Floras with a similar mixture are well displayed on the limestone pavements of Öland and Gotland, where Iversen (1954) notes the survival of a rich 'late-glacial' flora.

Following the forest clearances of the Neolithic at *c.* 5000 years ago and the agricultural revolution which followed, many light-demanding species spread to the cultivated areas and associated wasteland habitats, as shown by the number of species common to such areas today and the stadial flora (Table 11.4) and considered weeds of cultivated ground. Godwin (1975) has described these changes in detail. If the climate reverts to a cold stage in future, these plants, with their genetic diversity, are likely to be again a significant component of the flora. As this history of the cold stage flora shows, they are able to respond relatively rapidly to environmental changes, an ability reflected in the widespread distributions commonly seen.

A historical explanation of the origin of disjunct distributions of British species of open habitats has been fully considered by Pigott & Walters (1954), connecting the Devensian late-glacial conditions to disjunct areas of the kind listed above where open ground species could survive the spread of forests in the post-glacial. Habitats are listed, as are the species associated with them and the present distribution categories of the species. Since that time, much more has been discovered of the stadial floras, and this new knowledge strengthens the interpretations of Pigott & Walters (1954). Thus, many of the disjunct species are continental in distribution, which conforms with the continental nature of the stadial climates discussed previously. Their survival in disjunct areas of suitable post-glacial habitat is the survival of species of the stadial flora, a Gleasonian concept.

Two particular cases where a connection between Devensian floras and present floras has been made illustrate these matters. The first is the investigation by Turner *et al.* (1973) of the vegetational history of the area of upland Teesdale characterised by a well-known flora, long regarded as including relict species related to the last cold stage flora. It is remarkable that many species of late-glacial type are recorded at several times in the post-glacial vegetational pollen record, indicating their survival through a period of open woodland (the period of forset maximum) in the post-glacial, so confirming their relict status. These species include a number of species rare in the area, including *Armeria maritima*, *Dryas octopetala*, *Helianthemum* and *Plantago maritima*. The second is a vegetational record from an area of chalk grassland in the Yorkshire Wolds. Here Bush & Flenley (1987) have described a sequence from Devensian late-glacial to the Flandrian forested period where grassland pollen assemblages with characteristic stadial taxa (*Helianthemum, Campanula, Plantago*) appear to have survived through to 8000 BP at least. It is possible that chalk grassland may have been a continuous, if very local, community through the post-glacial. If so, it is a transformed survivor of stadial grassland.

This overall view shows the cold stage stadial flora to have a long and complex history, originating in the latest Tertiary, occupying a major part of Quaternary time in our area, and surviving short periods of forest dominance at times of climatic amelioration. It is not surprising that the taxonomy of the species concerned is very complex.

16
A final word

A question of terminology

The naming of segments of the Quaternary in terms of the sediment sequence (lithostratigraphy), or of the fossils or assemblages of fossils (biostratigraphy) or of time (identified via litho- or biostratigraphy and absolute dating – chronostratigraphy) is governed by national or international codes (e.g. Whittaker *et al.* 1991; Salvador 1994). Lowe & Walker (1997b) describe the concepts involved, which are clear but not always easily applicable to the complexities of the Quaternary geological record. A 'secondary' classification, important in the Quaternary, is the division of time into climatostratigraphic units, through interpretation of the climatic significance of the litho- and biostratigraphy. The cold and temperate stages of the Quaternary are such units. The naming of the stages is usually based on type sites which exemplify their climatic status. Stages and substages named in Chapter 2 are so defined.

Litho- and biostratigraphic evidence is used to define cold stages, but this interpretation of climate must be considered a broad approach to climatic conditions. For example, a particular glacial advance can result from annual or seasonal precipitation changes and/or from annual or seasonal temperatures changes. A particular vegetation formation may be associated with a similar variety of precipitation and temperature. Thus grassland can occur from the polar to the tropical regions. The problem is to define climates more accurately. The term temperate, used to define temperate stages or interstadials within cold stages, covers a wide variety of climates, from oceanic to continental, and a variety of vegetation formations, even within the area of the *Flora Europaea*. Changes in inferred annual temperature or July temperatures, with no reference to possibly very significant seasonal changes, may be little guide to climatic conditions.

These difficulties are discussed in Chapter 14 in relation to climate interpretation of the flora and fauna. They are compounded by the necessary habit of using present-day vegetation formations and their associated climates as a key to interpreting past formations and climates, such as describing particular floras as 'arctic'. These present-day units of biota are the biomes, characterised by their flora, fauna and climate, such as tundra, steppe, deciduous forest. They are an expression of present conditions of climate and of the results of past events governing migration and speciation, with their internal diversity conditioned by geology and landforms.

In the region we have been considering, the idea that such a biome, with its characteristic flora, has a long history in the Quaternary is a Clementsian concept, perhaps with an element of uniformitarianism. The alternative Gleasonian view regards as important the responses of individuals (species and subspecific ranks) to changing environments. As seasonal temperatures and precipitation change, plants will respond according to their ecological requirements and biotype content, both of which may be relatively unknown in the living species.

Comparisons of fossil assemblages are often made with arctic formations such as polar desert or tundra, or with steppe. But the climates of polar desert and tundra cannot compare with those of cold stages at our latitudes, though the more southern formation of steppe may bear comparison. As we have discussed, the cold stage flora cannot be compared with present northern floras, a correlative of the difference between present northern climates and cold stage climates.

Iversen (in Degerbøl & Iversen 1945) clearly stated this point in his conclusion that Danish Weichselian late-glacial vegetation differed in important points from that of the present Arctic:

it may therefore be questioned whether one is justified in using the appellation Arctic or sub-Arctic of late-glacial vegetation. The points of similarity with the Central-European alpine vegetation are in reality more prominent . . . but the steppe elements were never so prominent as to provide reasons for preferring the term 'late-glacial steppe' to 'late-glacial tundra'.

Iversen suggested the term 'park-tundra' for

the curious open country that spread over immense regions at the close of the glacial age, and with its mixture of grassland and tree-islands provided ideal living conditions for the rich late-glacial fauna.

This vegetation was the direct descendant of the cold stage stadial vegetation we have described.

In a historical context, it is interesting to note that Nehring (1890), more concerned with fauna than flora, devoted much thought to similar

A final word

questions. He compared the cold stage conditions to those of subarctic steppe or similar in eastern Russia and western Siberia. He considered the use of the term 'park-steppe' (cf. Iversen, above), but preferred 'wald-insel steppe' because he thought, understandably, that 'park' was unsuitable as it suggested an English landscape with an oceanic climate.

These questions derive from attempts to describe cold stage biota in terms of present biomes. They are distinct from stratigraphical terminology relying on particular fossils (e.g the Weichselian Late-glacial Oldest, Older or Younger *Dryas* periods), those based on pollen assemblage biozones (Devensian woodland and herb biozones of Phillips (1976)), and those based on type sites (Weichselian Late-glacial Bølling, Allerød periods), all referring to the fossil record. The problem lies not only in translating occurrences of fossils into climatic terms, a question discussed in Chapter 14. It lies also in translating fossil assemblages into vegetation formations which have climatic significance. From the fossil record it is clear that in our region no present-day equivalents are known of certain cold stage stadial assemblages, and therefore no term(s) exists for describing the formation(s) they represent.

The concept of steppe-tundra attempts to overcome this descriptive problem, but in doing so confuses the climatic meaning of the stadial assemblages. Relict areas with floristically steppe-tundra associations are known in northern regions (Alaska, Siberia; see Chapter 12), indicating the assemblages at present require low annual temperatures with suitable summer warmth. This is the kind of formation which must have existed much more widely at lower latitudes in stadial time, under different, but perhaps seasonally similar, climatic conditions, giving rise to a biome unknown widely today. Terms are needed to describe these formations, terms which do not rely on present formations with their own climatic connotations. Such terms should describe the physiognomy of the vegetation and the major floristic components, for example, stadial herb-rich grassland, rather than making the direct relation to present-day formations.

A survey of the analyses

Chapter 1 proposed various aims for the analysis of the cold stage flora, and the achievement of these aims may now be considered. The data tables show which taxa are found where, when and in what quantity, with Chapter 7 summarising records family-by-family. The translation of quantity to actual representation in the vegetation is a much more difficult matter, discussed in Chapters 8 and 12. Details of the representation of taxa by macroscopic remains and/or pollen are also given in Chapter 8. The advantage of macroscopic remains for species identification is strikingly illustrated. With a large number of species identified, it is possible to analyse the flora in terms of present distribution (Chapter 11). The results show that most species are known in the present British flora. They clarify the relation between the cold stage flora and the present flora, discussed in Chapter 15. An analysis of the distribution in time of the species shows a large number have been recorded in temperate and cold stages, indicating the ability of the Linnean species to survive climatic changes.

Aspects of the biology of the flora are described in Chapter 9. The life forms, life spans and variability of the species identified are considered, revealing difference and similarities with present northern floras. In relation to the vegetation, Chapter 10 discusses the habitats of the cold stages, and Chapter 12 the vegetation and flora. Identification of particular communities is difficult, and the plant records have been interpreted in terms of types of vegetation (in the sense of Tansley (1911)). Knowledge of the diversity of habitats and vegetation is essential to the understanding of the cold stage flora. Regional grassland is indicated by the stadial assemblages, contrasting with forest development in the Early Devensian Chelford and Brimpton Interstadials.

The taphonomy of the fossil assemblages is discussed in Chapter 3, with the variety of sedimentary environments in cold stages considered. The problem of reworking of fossils, especially temperate stage fossils into cold stage assemblages, is a major difficulty for the interpretation of the assemblages of macroscopic remains and pollen.

The environmental significance of the assemblages is discussed in terms of habitat (Chapter 10) and climate (Chapter 13). The latter is complex, relying on data about the relation between present distribution and climate. Evidence from geology, vegetation and species is considered, also the differences in climatic indication by plants and Coleoptera. An explanation is sought in terms of continental seasonal climates and diversity of habitats. A particular problem is the extent to which climatic change within a cold stage can be identified from fossil assemblages, especially because the widespread grassland can occur in a variety of climates.

Cold stage biota (vertebrate, Coleopteran and molluscan faunas) are considered in relation to the evidence from the flora in Chapter 14. Non-analogue faunas are present, as with the flora. The origin and the fate of the flora is summarised in Chapter 15.

Reviewing the results of the analysis of the cold stage flora, the contributions made by records of pollen and macroscopic remains are clearly different and complementary. With the wide dispersal of pollen, the pollen records are of prime importance for characterising the regional vegetation, even though there may be significance from time to time of more locally derived pollen. The identifications of the most abundant taxa are often at family or generic level, and therefore have a less precise meaning in terms of environment. Macroscopic remains, more locally derived and allowing specific identifications, give information on local habitats and environments in the catchment. The large number of species identified from macroscopic remains leads to a much better assessment of habitats, vegetation and climate and of the relation of the flora to the present-day flora. There is a huge loss of potential if only pollen is studied, as often has to be the case in analyses of long sequences in deep lakes.

While the aims of the analysis have been largely achieved, it is important not to forget the many limitations. Specific identifications have been taken as published, but often identification is more difficult than would appear from the taxa lists. The indications of habitat and climate are similarly difficult because of lack of knowledge of the ecology of species. Even with these limitations, however, the mass of plant records has been found to be informative.

The future

While the general nature of the cold stage flora is well established, the sites yielding plant remains are very unequally distributed, with the concentration in the south and east of Britain. As a result it is not possible to distinguish clearly regional variations in flora and environments. The Scottish Devensian site of Sourlie (H. Bos & J.H. Dickson, unpubl.), with both pollen and macro records, is the most informative of the northern sites. The Cornish sites of Scourse (1991), yielding pollen diagrams indicating grassland, are the only sites in the south-west, and there are few sites in the west of the British Isles. Identification of regional variation in the western region of the continent of Europe would be very interesting in relation to the variation of cold stage climates as well as flora and vegetation. But the paucity of sites and the confusing effects of taphonomy, so significant in controlling the constitution of the assemblages, will make such identification difficult, as well as hindering the identification of climatic changes within cold stages.

As well as the limitations imposed by the distribution of sites, the variability of many cold stage species also hinders interpretations of the assemblages. Their taxonomy is difficult and reference collections may be incomplete. Improvements in taxonomy and the study of variable species by modern methods such as isozyme or DNA analysis (Briggs & Walters 1997), particularly if they are applicable to fossil macroscopic remains as well as living plants, would be of very great interest in terms of the evolution of species under conditions of climatic change. The development of atlases of macroscopic plant parts would greatly aid this aspect of palaeoecology.

The database given here (to 1995) can be expanded as new fossil assemblages are described, as taxonomy and identification improves, as dating is improved and as stratigraphy is revised. The present database and the interpretation of cold stage floras discussed here will hopefully form a sound basis for these developments.

References

Adam, P. (1977). The ecological significance of 'halophytes' in the Devensian flora. *New Phytologist*, **78**, 237–244.

Agriculture Canada Expert Committee on Soil Survey (1987). *The Canadian system of soil classification*. 2nd Ed. Agriculture Canada Publication 1646, 164 pp.

Andersen, J. (1993). Beetle remains as indicators of the climate in the Quaternary. *Journal of Biogeography*, **20**, 557–562.

Andersen, J. (1996). Do beetle remains reliably reflect the macroclimate in the past ? – a reply to Coope and Lemdahl. *Journal of Biogeography*, **23**, 120–121.

Andersen, S.T. (1961). Vegetation and its environment in Denmark in the Early Weichselian Glacial (Last Glacial). *Danmarks Geologiske Undersøgelse*, II Raekke, No. 75.

Andersen, S.T., de Vries, Hl. & Zagwijn, W.H. (1960). Climatic change and radiocarbon dating in the Weichselian Glacial of Denmark and The Netherlands. *Geologie en Mijnbouw*, **39**, 38– 42.

Andreev, V.N. & Aleksandrova, V.D. (1981). Geobotanical division of the Soviet Arctic. In *Tundra ecosystems: a comparative analysis*, ed. L.C. Bliss, O.W. Heal & J.J. Moore, pp. 25–37. Cambridge University Press.

Andrew, R. (1970). The Cambridge pollen reference collection. In *Studies in the vegetational history of the British Isles*, ed. D. Walker & R.G. West, pp. 225–231. Cambridge University Press.

Andrew, R. & West, R.G. (1977). Pollen analyses from Four Ashes, Worcs. Appendix, pp. 242–246, to West, R.G. (1977), Early and Middle Devensian flora and vegetation. *Philosophical Transactions of the Royal Society of London, B*, **280**, 229–246.

Ashworth, A.C. (1977). A late Wisconsinan Coleopterous assemblage from southern Ontario and its environmental significance. *Canadian Journal of Earth Sciences*, **14**, 1625–1634.

Atkinson, T.C., Briffa, K.R.& Coope, G.R. (1987). Seasonal temperatures in Britain during the past 22,000 years, reconstructed using beetle remains. *Nature*, **325**, 587–592.

Aurivillius, C. (1883). Insect life in Arctic lands. In *Studier och Förskningar Foranledda af mina Resor i Höga Norden*, ed. A.E. Nordenskiöld, pp. 403–459. Stockholm: F. & J. Beijers Forlag.

Ballantyne, C.K. & Harris, C. (1994). *The periglaciation of Great Britain*. Cambridge University Press.

Batzli, G.O. (ed.) (1980). Patterns of vegetation and herbivory in arctic tundra environments. *Arctic and Alpine Research*, **12**, 401–578.

Bazzaz, F.A. (1996). *Plants in changing environments*. Cambridge University Press.

Behre, K.-E. (1989). Biostratigraphy of the last glacial period in Europe. *Quaternary Science Reviews*, **8**, 25–44.

Behre, K.-E. & van der Plicht, J. (1992). Towards an absolute chronology for the last glacial period in Europe: radiocarbon dates from Oerel, northern Germany. *Vegetation History and Archaeobotany*, **1**, 111–117.

Bell, F.G. (1968). *Weichselian glacial floras in Britain*. Ph.D. Thesis, University of Cambridge.

Bell, F.G. (1969). The occurrence of southern, steppe, and halophyte elements in Weichselian (Last-Glacial) floras from southern Britain. *New Phytologist*, **68**, 913–922.

Bell, F.G. (1970). Late Pleistocene floras from Earith, Huntingdonshire. *Philosophical Transactions of the Royal Society of London, B*, **258**, 347–378.

Berger, A., Imbrie, J., Hays, J., Kukla, G. & Saltzman, B. (eds.) (1984). *Milankovitch and climate*. Dordrecht: D. Riedel Publishing Company.

Berglund, B.E. (ed.) (1986). *Handbook of Holocene palaeoecology and palaeohydrology*. Chichester: Wiley.

Berglund, B.E., Birks, H.J.B., Ralska-Jasiewiczowa, M. & Wright, H.E. (eds) (1996). *Palaeoecological events during the last 15 000 years*. New York: Wiley.

Billings, W.D. & Mooney, H.A. (1968). The ecology of arctic and alpine plants. *Biological Reviews*, **43**, 481–529.

Birks, H.J.B. (1973). *Past and present vegetation of the Isle of Skye*. Cambridge University Press.

Birks, H.J.B. (1996). Great Britain – Scotland. In *Palaeoecological events during the last 15 000 years*, ed. B.E. Berglund *et al.*, pp. 95–143. New York: Wiley.

Birks, H.J.B. & Birks, H.H. (1980). *Quaternary palaeoecology*. London: Edward Arnold.

Birks, H.J.B. & Deacon, J. (1973). A numerical analysis of the past and present flora of the British Isles. *New Phytologist*, **72**, 877–902.

Bliss, L.C. (1971). Arctic and Alpine plant life cycles. *Annual Reviews of Ecology and Systematics*, 1971, 405–438.

Bliss, L.C. (1975). Tundra, grasslands, herblands, and shrublands and the role of herbivores. *Geoscience and Man*, **10**, 51–79.

Bliss, L.C. (1981a). North American and Scandinavian tundras and polar deserts. In *Tundra ecosystems: a comparative analysis*, ed. L.C. Bliss, O.W. Heal & J.J. Moore, pp. 8–24. Cambridge University Press.

Bliss, L.C. (1981b). Evolution of tundra ecosystems. In *Tundra ecosystems: a comparative analysis*, ed. L.C. Bliss, O.W. Heal & J.J. Moore, pp. 5–7. Cambridge University Press.

Boreham, S. & West, R.G. (1993). Late Quaternary sediments with *Chara* encrustations in southern Fenland. *Geological Magazine*, **130**, 743–744.

Bowen, D.Q. (1994). The Pleistocene of North West Europe. *Science Progress*, **76**, 209–223.

Bradley, R.S. (1985). *Quaternary palaeoclimatology*. London: Allen & Unwin.

Bradshaw, A.D. & McNeilly, T. (1991). Evolutionary response to global climatic change. *Annals of Botany*, **67**, 5–14.

Brandon, L.V. (1965). Groundwater hydrology and water supply in the District of Mackenzie, Yukon Territory and adjoining parts of British Columbia. *Geological Survey of Canada Paper*, 64–39.

Briggs, D. & Walters, S.M. (1997). *Plant variation and evolution*. 3rd Ed. Cambridge University Press.

Brochmann, C., Soltis, P.S. & Soltis, D.E. (1992). Recurrent formation and polyphyly of Nordic polyploids in *Draba* (Brassicaceae). *American Journal of Botany*, **79**, 673–688.

Brush, G.S. & Brush, L.M. (1994). Transport and deposition of pollen in an estuary: signature of the landscape. In *Sedimentation of organic particles*, ed. A. Traverse, pp. 33– 46. Cambridge University Press.

Bryant, I.D. (1983a). Facies sequence associated with some braided river deposits of Late Pleistocene age from Southern Britain. *Special Publications of the International Association of Sedimentologists*, **6**, 267–275.

Bryant, I.D. (1983b). The utilisation of arctic river analogues in the interpretation of periglacial river sediments from southern Britain. In *Background to palaeohydrology*, ed. K.J. Gregory, pp. 413–436. London: Wiley.

Bryant, I.D., Holyoak, D.T. & Moseley, K.A. (1983). Late Pleistocene deposits at Brimpton, Berkshire, England. *Proceedings of the Geologists' Association*, **94**, 321–343.

Burn, C.R. (1997). Cryostratigraphy, palaeogeography, and climate change during the early Holocene warm interval, western Arctic coast, Canada. *Canadian Journal of Earth Sciences*, **34**, 912–915.

Bush, M.B. & Flenley, J.R. (1987). The age of the British chalk grassland. *Nature*, **329**, 434–436.

Cajander, A.K. (1903). Beiträge zur Kenntnis der Vegetation der Alluvionem des nördlichen Eurasiens. 1. Die Alluvionen des unteren Lena-thales. *Acta Societatis Scientarum Fennicae*, **32**, 1–182.

Chambers, F.M. (1996). Great Britain – Wales. In *Palaeoecological events during the last 15 000 years*, ed. B.E. Berglund *et al.*, pp. 77–94. New York: Wiley.

Chapin, F.S. & Shaver, G.R. (1985). Arctic. In *Physiological ecology of North American plant communities*, ed. B.F. Chabot & H.A. Mooney, pp. 16–40. London: Chapman & Hall.

Chernov, Yu.I. (1985). *The living tundra*. Cambridge University Press.

Clapham, A.R. Tutin, T.G. & Warburg, E.F. (1962). *Flora of the British Isles*. 2nd Ed. Cambridge University Press.

Clapham, A.R., Tutin, T.G. & Moore, D.M. (1989). *Flora of the British Isles*. 3rd Ed. Cambridge University Press.

Colinvaux, P.A. (1986). Plain thinking on Bering Land Bridge Vegetation and Mammoth Populations. *Quarterly Review of Archaeology*, **7**, 8–9.

Coombe, D.E. (1994). 'Maritime' plants of roads in Cambridgeshire (v.c.29). *Nature in Cambridgeshire*, **36**, 37–60.

Coope, G.R. (1959). A Late Pleistocene insect fauna from Chelford, Cheshire. *Proceedings of the Royal Society of London, B*, **151**, 70–86.

Coope, G.R. (1968). Coleoptera from the 'Arctic Bed' at Barnwell Station, Cambridge. *Geological Magazine*, **105**, 482–486.

Coope, G.R. (1973). Tibetan species of dung beetle from Late Pleistocene deposits in Britain. *Nature*, **245**, 335–336.

Coope, G.R. (1974). Report on the Coleoptera from Wretton. *Philosophical Transactions of the Royal Society of London, B*, **267**, 414–418.

Coope, G.R. (1975). Climatic fluctuations in northwest Europe since the Last Interglacial, indicated by fossil assemblages of Coleoptera. *Geological Journal*, Special Issue No. 6, 153– 168.

Coope, G.R. (1977). Fossil coleopteran assemblages as sensitive indicators of climatic changes during the Devensian (Last) cold stage. *Philosophical Transactions of the Royal Society of London, B*, **280**, 313–340.

Coope, G.R. & Angus, R.B. (1975). An ecological study of a temperate interlude in the middle of the last glaciation, based on fossil Coleoptera from Isleworth, Middlesex. *Journal of Animal Ecology*, **44**, 365–391.

Coope, G.R., Morgan, A. & Osborne, P.J. (1971). Fossil Coleoptera as indicators of climatic fluctuations during the last glaciation in Britain. *Palaeogeography, Palaeoclimatology, Palaeoecology*, **10**, 87–101.

Coope, G.R., Gibbard, P.L., Hall, A.R., Preece, R.C., Robinson, J.E. & Sutcliffe, A.J. (1997). Climatic and environmental reconstructions based on fossil assemblages form Middle Devensian (Weichselian) deposits of the River Thames at South Kensington, central London, UK. *Quaternary Science Reviews*, **16**, 1163–1195.

Coope, G.R., Shotton, F.W. & Strachan, I. (1961). A Late Pleistocene fauna and flora from Upton Warren, Worcestershire. *Philosophical Transactions of the Royal Society of London*, **244**, 379–421.

Coupland, R.T. (1979). The nature of grassland. In *Grassland ecosystems of the world: analysis of grasslands and their uses*, ed. R.T. Coupland, pp. 23–29. Cambridge University Press.

Coxon, P. (1993). Irish Pleistocene biostratigraphy. *Irish Journal of Earth Sciences*, **12**, 83–105.

Coxon, P., Hall, A.R., Lister, A. & Stuart, A.J. (1980). New evidence on the vertebrate fauna, stratigraphy and palaeobotany of the interglacial deposits at Swanton Morley, Norfolk. *Geological Magazine*, **117**, 525–546.

Crawford, R.M.M. (1997). Habitat fragility as an aid to long-term survival in Arctic vegetation. In *Ecology of Arctic environments*, ed. S.J. Woodin & M. Marquiss, pp. 113–136. Oxford: Blackwell.

Crawford, R.M.M. & Abbott, R.J. (1994). Pre-adaptation of Arctic plants to climatic change. *Botanica Acta*, **107**, 271–278.

Crawford, R.M.M., Chapman, H.M. & Smith, L.C. (1995). Adaptation to variation in growing season length in Arctic populations of *Saxifraga oppositifolia* L. *Botanical Journal of Scotland*, **47**, 177–192.

Currant, A. & Jacobi, R. (1997). Vertebrate faunas of the British Late Pleistocene and the chronology of human settlement. *Quaternary Newsletter*, No. 82, 1–8.

Danks, H.V. (1990). Arctic insects: instructive diversity. In *Canada's missing dimension*, ed. C.R. Harington, pp. 444–470. Ottawa: Canadian Museum of Nature.

Darwin, C. (1859). *On the Origin of Species by means of Natural Selection*. Last (6th) Edition. London: Watts & Co.

Davis, A.J., Jenkinson, L.S., Lawton, J.H., Shorrocks, B. & Wood, S. (1998). Making mistakes when predicting shifts in species range in response to global warming. *Nature*, **391**, 783–786.

de Beaulieu, J.-L. (1996). Foreword. In *Palaeoecological events during the last 15000 years*, ed. B.E. Berglund *et al.*, pp. xix–xx. New York: Wiley.

de Beaulieu, J.-L. & Reille, M. (1992). The last climatic cycle at Grande Pile (Vosges, France): a new pollen profile. *Quaternary Science Reviews*, **11**, 431–438.

Degerbøl, M. & Iversen, J. (1945). The Bison in Denmark. *Danmarks Geologiske Undersøgelse*, II Raekke, No. 73.

Dickson, C.A. (1970). The study of plant macrofossils in British Quaternary deposits. In *Studies in the vegetational history of the British Isles*, ed. D. Walker & R.G. West, pp. 233–254. Cambridge University Press.

Dickson, J.H. (1973). *Bryophytes of the Pleistocene*. Cambridge University Press.

Donner, J.J. (1995). *The Quaternary history of Scandinavia*. Cambridge University Press.

Edlund, S.A. (1987). Plants: living weather stations. *Geos*, **16**, 9–13.

Ellenberg, H. (1974). Zeigerwerte der Gefässpflanzen Mitteleuropas. *Scripta Geobotanica*, **9**, 1–97.

Emiliani, C. (1955). Pleistocene temperatures. *Journal of Geology*, **63**, 538–578.

Emontspohl, A.-F. (1995). The northwest European vegetation at the beginning of the Weichselian glacial (Brørup and Odderade interstadials) – new data for northern France. *Review of Palynology and Palaeobotany*, **85**, 231–242.

Fitter, A.H. & Hay, R.K.M. (1987). *Environmental physiology of plants*. London: Academic Press.

Fitter, A.H. & Peat, H.J. (1994). The Ecological Flora Database. *Journal of Ecology*, **82**, 415–425.

Florschütz, F. (1958). Steppen und Salzumpfelementen aus dem Flora der letzen und vorletzen Eiszeit in den Niederland. *Flora*, **146**, 489–492.

Forbes, E. (1846). On the connexion between the Distribution of the existing fauna and flora of the British Isles, and the Geological Changes which have affected their area, especially during the epoch of the Northern Drift. *Memoirs of the Geological Survey of Great Britain and of the Museum of Economic Geology in London*, **1**, 336–432.

Fox, J.F. (1981). Intermediate levels of soil disturbance maximise alpine plant diversity. *Nature*, **293**, 564–565.

French, H.M. (1976). *The periglacial environment*. London: Longman.

French, N.R. (1979). Introduction. In *Grassland ecosystems of the world: analysis of grasslands and their uses*, ed. R.T. Coupland, pp. 41–48. Cambridge University Press.

Funder, S. & Ábrahamsen, N. (1988). Palynology in a polar desert, eastern North Greenland. *Boreas*, **17**, 195–207.

Funnell, B.M. (1995). Global sea level and the (pen-)insularity of late Cenozoic Britain. In *Island Britain: a Quaternary perspective*, ed. R.C. Preece, pp. 3–13. Geological Society Special Publication No. 96.

Gibbard, P.L. (1985). *The Pleistocene history of the Middle Thames Valley*. Cambridge University Press.

Gibbard, P.L., West, R.G., Andrew, R. & Pettit, M. (1992). The margin of a Middle Pleistocene ice advance at Tottenhill, Norfolk, England. *Geological Magazine*, **129**, 59–76.

Gibbard, P.L., West, R.G., Zagwijn, W.H., Balson, P.S., Burger, A.W., Funnell, B.M., Jeffery, D.H., de Jong, J., van Kolfschoten, T., Lister, A.M., Meijer, T., Norton, P.E.P., Preece, R.C., Rose, J., Stuart, A.J., Whiteman, C.A. & Zalasiewicz, J.A. (1991). Early and Middle Pleistocene correlations in the southern North Sea basin. *Quaternary Science Reviews*, **10**, 23–52.

Giterman, R.E. (1985). Kolyma lowland vegetation in the Pleistocene cold epochs and the problem of polar Beringia landscapes. In *Beringia in the Cenozoic Era*, ed. V.L. Kontrimavichus, pp. 214–220. Rotterdam: Balkema.

Glück, H. (ed.) (1936). Die Süsswasser-Flora Mitteleuropas: Heft 15, *Pteridophyten und Phanerogamen*. Jena: Gustav Fischer.

Godwin, H. (1964). Late-Weichselian conditions in south eastern Britain: organic deposits at Colney Heath, Herts. *Proceedings of the Royal Society of London, B*, **160**, 258–275.

Godwin, H. (1968). The development of Quaternary palynology in the British Isles. *Review of Palaeobotany and Palynology*, **6**, 9–20.

Godwin, H. (1975). *The history of the British Flora*. 2nd Ed. Cambridge University Press.

Graham, R.W. & the FAUNMAP Working Group (1996). Spatial response of mammals to Late Quaternary environmental fluctuations. *Science*, **272**, 1601–1606.

Greig, J. (1996). Great Britain – England. In *Palaeoecological events during the last 15 000 years*, ed. B.E. Berglund *et al.*, pp. 15–76. New York: Wiley.

Grichuk, V.P. (1984). Late Pleistocene vegetation history. In *Late Quaternary environments of the Soviet Union*, ed. A.A. Velichko, H.E. Wright & C.W. Barnosky, pp. 155–178. London: Longman.

Grichuk, M.P. & Grichuk, V.P. (1960). O prilednikovoi rastitelnosti na territorii S.S.S.R. In *Periglacial phenomena on the territory of the U.S.S.R.*, ed. K.K. Markov & A.I. Popov, pp. 66–100. Moscow University Press.

Griffiths, M.E. & Proctor, M.C.F. (1956). *Helianthemum canum* (L.) Baumg. *Journal of Ecology*, **44**, 677–682.

Grime, J.P., Hodgson, J.G. & Hunt, R. (1990). *The abridged comparative plant ecology*. London: Chapman & Hall.

Guiot, J. (1990). Methodology of palaeoclimatic reconstruction from pollen in France. *Palaeogeography, Palaeoclimatology, Palaeoecology*, **80**, 49–69.

Guiot, J., de Beaulieu, J.-L., Cheddadi, R., David, F., Ponel, P. & Reille, M. (1993). The climate in western Europe during the last Glacial/Interglacial cycle derived from pollen and insect remains. *Palaeogeography, Palaeoclimatology, Palaeoecology*, **103**, 73–93.

Guthrie, R.D. (1990). *Frozen fauna of the Mammoth Steppe*. London: University of Chicago Press.

Hadač, E. (1963). On the history and age of some arctic plant species. In *North Atlantic Biota and their history*, ed. Á. Löve & D. Löve, pp. 207–219. Oxford: Pergamon Press.

Hall, A.R. (1980). Late Pleistocene deposits at Wing, Rutland. *Philosophical Transactions of the Royal Society of London, B*, **289**, 135–164.

Harrison, R.G. (ed.) (1993). *Hybrid zones and the evolutionary process*. Oxford University Press.

Haviland, M.D. (1926). *Forest, steppe and tundra*. Cambridge University Press.

Havinga, A.J. (1984). A 20–year experimental investigation into the differential corrosion susceptibility of pollen and spores in various soil types. *Pollen et Spores*, **26**, 541–558.

Hibbert, D. (1982). History of the Steppe-Tundra concept. In *Palaeoecology of Beringia*, ed. D.M. Hopkins *et al.*, pp. 153–156. London: Academic Press.

Holyoak, D.T. (1984). Taphonomy of prosective plant macrofossils in a river catchment on Spitsbergen. *New Phytologist*, **98**, 405–423.

Holyoak, D.T. & Preece, R.C. (1983). Evidence of a high Middle Pleistocene sea-level from estuarine deposits at Bembridge, Isle of Wight, England. *Proceedings of the Geologists' Association*, **94**, 231–244.

Hopkins, D.M, Matthews, J.V., Schweger, C.E & Young, S.B. (eds.) (1982). *Palaeoecology of Beringia*. London: Academic Press.

Hultén, E. (1937). *Outline of the history of arctic and boreal biota during the Quaternary Period*. Stockholm: Bokförlags Aktieboleget Thule.

Hultén, E. (1958). The Amphi-Atlantic plants and their phytogeographical connections. *Kungliga Svenska Vetenskapsakademiens Handlingar*, **7**, No. 1.

Hultén, E. (1962). The Circumpolar Plants. I. *Kungliga Svenska Vetenskapsakademiens Handlingar*, **8**, No. 5.

Hultén, E. (1971). *Atlas över växternas udbredning i Norden*. Stockholm: Generalstabens Litografiska anstalts förlag.

Huntley, B. (1990). Studying global change: the contribution of Quaternary palynology. *Palaeogeography, Palaeoclimatology, Palaeoecology (Global and Planetary Change Section)*, **82**, 53–61.

Huntley, B. (1993). The use of climatic response surfaces to reconstruct palaeoclimate from Quaternary pollen and plant macrofossil data. *Philosophical Transactions of the Royal Society of London, B*, **341**, 215–223.

Huntley, B. (1996). Quaternary palaeoecology and ecology. *Quaternary Science Reviews*, **15**, 591–606.

Huntley, B. & Birks, H.J.B. (1983). *An atlas of past and present pollen maps for Europe. 0–13 000 years ago*. Cambridge University Press.

Iversen, J. (1944). *Viscum, Hedera* and *Ilex* as climate indicators. A contribution to the study of the Post-Glacial temperature climate. *Geologiska Föreningens Stockholm, Förhandlingar*, **66**, 463–483.

Iversen, J. (1954). The Late-Glacial flora of Denmark and its relation to climate and soil. *Danmarks Geologiske Undersøgelse*, II Raekke, **80**, 86–119.

Iversen, J. (1973). The development of Denmark's Nature since the Last Glacial. *Danmarks Geologiske Undersøgelse*, V Raekke, No. 7–C.

Jackson, S.T. & Givens, C.R. (1994). Late Wisconsinan vegetation and environment of the Tunica Hills Region, Louisiana/Mississippi. *Quaternary Research*, **41**, 316–325.

Jermy, A.C. & Tutin, T.G. (1968). *British Sedges*. London: Botanical Society of the British Isles.

Jermy, A.C., Chater, A.O. & David, R.W. (1982). *Sedges of the British Isles*. 2nd Ed. London: Botanical Society of the British Isles.

Jessen, K. & Milthers, V. (1928). Stratigraphical and palaeontological studies of interglacial freshwater deposits in Jutland and north-west Germany. *Danmarks Geologiske Undersøgelse*, II Raekke, No. **48**

Kasse, C., Bohncke, S.J.P. & Vandenberghe, J. (1995). Fluvial periglacial environments during the Middle Weichselian in the northern Netherlands with special reference to the Hengelo Interstadial. *Mededelingen Rijks Geologische Dienst*, **52**, 387–413.

Kerney, M.P. (1963). Late-glacial deposits on the Chalk of south-east England. *Philosophical Transactions of the Royal Society of London, B*, **246**, 203–254.

Kerney, M.P. (1977). British Quaternary non-marine Mollusca: a brief review. In *British Quaternary Studies: recent advances*, ed. F.W. Shotton, pp. 31–42. Oxford: Clarendon Press.

Kerney, M.P., Preece, R.C. & Turner, C. (1980). Molluscan and plant biostratigraphy of some Late Devensian and Flandrian deposits in Kent. *Philosophical Transactions of the Royal Society of London, B*, **291**, 1–43.

Kolstrup, E. (1979). Herbs as July temperature indicators for parts of the Pleniglacial and Late-glacial in The Netherlands. *Geologie en Mijnbouw*, **58**, 377–380.

Kolstrup, E. (1980). Climate and stratigraphy in northwestern Europe between 30,000 b.p. and 13,000 b.p., with special reference to The Netherlands. *Mededelingen Rijks Geologische Dienst*, **32**, 181–238.

Kolstrup, E. (1990). The puzzle of Weichselian vegetation types poor in trees. *Geologie en Mijnbouw*, **69**, 253–262.

Kolstrup, E. (1995). Palaoenvironments in the north European lowlands between 50 and 10 ka BP. *Acta Zoologica Cracoviensia*, **38**, 35–44.

Kulczynski, S. (1923). Das Boreale und Arktische-Alpine Element in der Mittel-Europäischen Flora. *Bulletin international de l'Academie de Cracovie*, Ser. B (1923), 127–214.

Lamb, A.L. & Ballantyne, C.K. (1998). Palaeonunataks and the altitude of the last ice sheet in the SW Lake District, England. *Proceedings of the Geologists' Association*, **109**, 305–316.

Lambert, C.A., Pearson, R.G. & Sparks, B.W. (1963). A flora and fauna from

Late Pleistocene deposits at Sidgwick Avenue, Cambridge. *Proceedings of the Linnean Society of London*, **174**, 13–29.

Lawton, J.H. (1995). The response of insects to environmental change. In *Insects in a changing environment*, ed. R. Harrington & N.E. Stork, pp. 3–26. London: Academic Press.

Laxton, N.F., Burn, C.R. & Smith, C.A.S. (1996). Productivity of Loessal Grasslands in the Kluane Lake Region, Yukon Territory, and the Beringian 'Production Paradox'. *Arctic*, **49**, 129–140.

Lichti-Federovich, S. & Ritchie, J.C. (1968). Recent pollen assemblages from the western interior of Canada. *Review of Palaeobotany and Palynology*, **7**, 297–344.

Lowe, J.J. & Walker, M.J.C. (1997a). Temperature variations in NW Europe during the last Glacial/Interglacial transition (14–9 ^{14}C ka BP) based upon the analysis of Coleopteran assemblages – the contribution of Professor G.R. Coope. *Quaternary Proceedings*, No. 5, 165–175.

Lowe, J.J. & Walker, M.J.C. (1997b). *Reconstructing Quaternary environments.* 2nd Ed. London: Longman.

Maddy, D., Coope, G.R., Gibbard, P.L., Green, C.P. & Lewis, S.G. (1994). Reappraisal of Middle Pleistocene fluvial deposits near Brandon, Warwickshire, and their significance for the Wolston glacial sequence. *Journal of the Geological Society*, **151**, 221–223.

Major, J. (1980). Review of 'The Arctic Floristic Region', ed. B.A. Yurtsev. *Arctic and Alpine Research*, **12**, 581–583.

Mangerud, J. (1991). The Last Interglacial/Glacial cycle in Northern Europe. In *Quaternary landscapes*, ed. L.C.K. Shane & E.J. Cushing, pp. 38–75. Minneapolis: University of Minnesota Press.

Martinson, D.G., Pisias, N.G., Hays, J.D., Imbrie, J., Moore, T.C. & Shackleton, N.J. (1987). Age dating and the orbital theory of the ice ages: development of a high resolution 0–300,000 year chronostratigraphy. *Quaternary Research*, **27**, 1–29.

Matthews, J.R. (1937). Geographical relationships of the British Flora. *Journal of Ecology*, **25**, 1–90.

Matthews, J.R. (1955). *Origin and distribution of the British Flora.* London: Hutchinson.

Matthews, J.V. (1982). East Beringia in Late Wisconsin time: a review of the biotic evidence. In *Palaeoecology of Beringia*, ed. D.M. Hopkins *et al.*, pp. 127–152. London: Academic Press.

McGlone, M.S. & Moar, N.T. (1997). Pollen–vegetation relationships on the subantarctic Auckland Islands, New Zealand. *Review of Palaeobotany and Palynology*, **96**, 317–338.

Menke, B. (1975). Vegetationsgeschichte und Florenstratigraphie Nordwestdeutschlands im Pliozän und Frühquartär. Mit einem Beitrag zur Biostratigraphie des Weichselfrühglazials. *Geologisches Jahrbuch*, **A26**, 3–151.

Mitchell, F.J.G., Bradshaw, R.H.W., Hannon, G.E., O'Connell, M., Pilcher, J.R. & Watts, W.A. (1996). Ireland. In *Palaeoecological events during the last 15 000 years*, ed. B.E. Berglund *et al.*, pp. 1–13. New York: Wiley.

Mitchell, G.F., Penny, L.F., Shotton, F.W. & West, R.G. (1973). *A correlation of Quaternary deposits in the British Isles.* Geological Society of London, Special Report No. 4, 99 pp.

Mitchell, G.F. & Ryan, M. (1997). *Reading the Irish landscape.* Dublin: Town House.

Mooney, H.A. & Billings, W.D. (1961). Comparative physiological ecology of arctic and alpine populations of *Oxyria digyna*. *Ecological Monographs*, **31**, 1–29.

Nathorst, A.G. (1914). Neuere Erfahrungen von dem Vorkommen fossiler Glacialpflanzen und einiger darauf besonders für Mitteldeutschland basierte Schlussfolgerungen. *Geologiska Föreningens i Stockholm Förhandlingar*, **36**, 267–307.

Nehring, A. (1890). *Ueber Tundren und Steppen der Jetzt- und Vorzeit.* Berlin: Ferd. Dummlers Verlagsbuchhandlung.

Nilsson, T. (1983). *The Pleistocene. Geology and life in the Quaternary Ice Age.* London: D.Reidel.

Pennington, W. (1986). Lags in adjustment of vegetation to climate caused by the pace of soil development: evidence from Britain. *Vegetatio*, **67**, 105–118.

Pennington, W. (1996). Limnic sediments and the taphonomy of Lateglacial pollen assemblages. *Quaternary Science Reviews*, **15**, 501–520.

Perring, F.H. & Walters, S.M. (eds.) (1982). *Atlas of the British Flora.* 3rd Ed. Botanical Society of the British Isles.

Pettersson, B. (1965). Gotland and Öland. Two limestone islands compared. *Acta Phytogeographica Suecica*, **50**, 131–140.

Phillips, L. (1976). Pleistocene vegetational history and geology in Norfolk. *Philosophical Transactions of the Royal Society London, B*, **275**, 215–286.

Pielou, E.C. (1994). *A naturalist's guide to the Arctic.* London: University of Chicago Press.

Pigott, C.D. & Walters, S.M. (1954). On the interpretation of the discontinuous distributions shown by certain British species of open habitats. *Journal of Ecology*, **42**, 95–116.

Polunin, N. (1951). The real Arctic: suggestions for its limitation, subdivision and characterisation. *Journal of Ecology*, **39**, 308–315.

Polunin, N. (1959). *Circumpolar Arctic Flora.* Oxford: Clarendon Press.

Porsild, A.E. (1955). *The vascular plants of the Western Canadian Arctic Archipelago.* Bulletin 135, National Museum of Canada.

Porsild, A.E. & Cody, W.J. (1980). *Vascular plants of continental Northwest Territories, Canada.* Ottawa: National Museums of Canada.

Praeger, R.L. (1913). On the buoyancy of seeds of some Britannic plants. *Scientific Proceedings, Royal Dublin Society*, **14**, 13–62.

Preece, R.C. (1995). Edward Forbes (1815–1854) and Clement Reid 1853–1916): two generations of pioneering polymaths. *Archives of Natural History*, **22**, 419–435.

Preece, R.C. (1997). The spatial response of non-marine Mollusca to past climatic changes. In *Past and future rapid environmental changes: the spatial and evolutionary responses of terrestrial biota*, ed. B. Huntley *et al.*, pp. 163–177. Berlin: Springer Verlag.

Preston, C.D. (1995). *Pondweeds of Great Britain and Ireland.* B.S.B.I. Handbook No. 8. London: Botanical Society of the British Isles.

Preston, R.D. & Hill, M.O. (1997). The geographical relationships of British and Irish vascular plants. *Botanical Journal of the Linnean Society*, **124**, 1–120.

Printz, H. (1921). The vegetation of the Siberian-Mongolian Frontiers. *Contributiones ad Floram Asiae Interioris Pertinentes*, III. Trondheim: Det Kongelige Norske Videnskabers Selskab.

Ran, E.T.H. (1990). Dynamics of vegetation and environment during the Middle Pleniglacial in the Dinkel Valley (The Netherlands). *Mededelingen Rijks Geologische Dienst*, **44**, 141–205.

Ran, E.T.H. & van Huissteden, J. (1990). The Dinkel Valley in the Middle Pleniglacial: dynamics of a tundra river system. *Mededelingen Rijks Geologische Dienst*, **44**, 210–220.

Raunkiaer, C. (1934). *The life forms of plants and statistical plant geography.* Oxford: Clarendon Press.

Reid, C. (1899). *The origin of the British Flora.* London: Dulau & Co.

Reid, C. (1911). The relation of the present plant population of the British Isles to the Glacial Period. *Irish Naturalist*, **20**, 201–209.

Rieseberg, L.H. & Wendel, J.F. (1993). Introgression and its consequences in plants. In *Hybrid zones and the evolutionary process*, ed. R.G. Harrison, pp. 70–109. Oxford University Press.

Ritchie, J.C. (1987). *Postglacial vegetation of Canada.* Cambridge University Press.

Ritchie, J.C., Hadden, K.A. & Gajewski, K. (1987). Modern pollen spectra from lakes in arctic western Canada. *Canadian Journal of Botany*, **65**, 1605–1613.

Ritchie, J.C. & Cwynar, L.C. (1982). The Late Quaternary vegetation of the North Yukon. In *Palaeoecology of Beringia*, ed. D.M. Hopkins *et al.*, pp. 113–126. London: Academic Press.

Ritchie, J.C. & Lichti-Federovich, S. (1967). Pollen dispersal phenomena in Arctic–Subarctic Canada. *Review of Palaeobotany and Palynology*, **3**, 255–266.

Salisbury, E.J. (1932). The East Anglian Flora. *Transactions of the Norfolk and Norwich Naturalists' Society*, **13**, 191–263.

Salisbury, E.J. (1942). *The reproductive capacity of plants.* London: Bell & Sons.

Salvador, A. (ed.) (1994). *International Stratigraphic Guide: a guide to stratigraphic classification, terminology and procedure.* 2nd Ed. International Subcommission on Stratigraphy, Geological Society of America.

Schouw, J.F. (1822). *Grundtraek til en almindelig Plantegeographie.* Copenhagen: Gyldendal.

Schweger, C.E. (1982). Late Pleistocene vegetation of Eastern Beringia: pollen analysis of dated alluvium. In *Palaeoecology of Beringia*, ed. D.M. Hopkins *et al.*, pp. 95–112. London: Academic Press.

Schweger, C.E. (1990). The full-glacial ecosystem of Beringia. *Prehistoric Mongoloid Dispersals (University of Tokyo)*, No. 7, 35–51.

Scourse, J.D. (1985). *Late Pleistocene stratigraphy of the Isles of Scilly and adjoining regions.* Ph.D. Thesis, University of Cambridge.

Scourse, J.D. (1991). Late Pleistocene stratigraphy and palaeobotany of the Isles of Scilly. *Philosophical Transactions of the Royal Society of London, B*, **334**, 405–448.

Seddon, M.B. & Holyoak, D.T. (1985). Evidence of sustained regional permafrost during deposition of fossiliferous Late Pleistocene river sediments at Stanton Harcourt (Oxfordshire, England). *Proceedings of the Geologists' Association*, **96**, 53–71.

Seret, G., Guiot, J., Wansard, G., de Beaulieu, J.-L. & Reille, M. (1992). Tentative palaeoclimate reconstruction linking pollen and sedimentology in La Grande Pile (Vosges, France). *Quaternary Science Reviews*, **11**, 425–430.

Seward, A.C. (1935). Discussion on the origin and relationship of the British Flora. *Proceedings of the Royal Society of London, B*, **118**, 197–241.

Shackleton, N.J & Opdyke, N.D. (1973). Oxygen isotope and palaeomagnetic stratigraphy of equatorial Pacific core v28–238: Oxygen isotope temperatures and ice volumes on a 10^5 and 10^6 year scale. *Quaternary Research*, **3**, 39–55.

Sheard, J.W. & Geale, D.W. (1983). Vegetation studies at Polar Bear Pass,

Bathurst Island, N.W.T. 1. Classifiaction of plant communities. *Canadian Journal of Botany*, **61**, 1618–1636.

Sher, A.V. (1997). Late Quaternary extinction of large mammals in northern Eurasia: A new look at the Siberian contribution. In *Past and future rapid environmental changes: the spatial and evolutionary responses of terrestrial biota*, ed. B. Huntley *et al.*, pp. 319–339. Berlin: Springer Verlag.

Simpson, I.M. & West, R.G. (1958). On the stratigraphy and palaeobotany of a late-Pleistocene organic deposit at Chelford, Cheshire. *New Phytologist*, **57**, 239–250.

Smith, C.A.S., Kennedy, C.E., Hargrave, A.E. & McKenna, K.M. (1989). Soil and vegetation survey of Herschel Island, Yukon Territory. *Yukon Territory Soil Survey Report No. 1*. LRRC Contribution No. 88–26. Agriculture Canada, Whitehorse, Yukon.

Sparks, B.W. & West, R.G. (1959). The palaeoecology of the interglacial deposits at Histon Road, Cambridge. *Eiszeitalter und Gegenwart*, **10**, 123–143.

Sparks, B.W. & West, R.G. (1972). *The Ice Age in Britain*. London: Methuen.

Stuart, A.J. (1982). *Pleistocene vertebrates in the British Isles*. London: Longman.

Stuart, A.J. (1995). Insularity and Quaternary vertebrate faunas in Britain and Ireland. In *Island Britain: a Quaternary perspective*, ed. R.C. Preece, pp. 111–125. Geological Society of London.

Sultan, S.E., Wilczek, A.M., Hann, S.D. & Brosi, B.J. (1998). Contrasting ecological breadth of co-occurring annual *Polygonum* species. *Journal of Ecology*, **86**, 363–383.

Szafer, W. (1954). Pliocene flora from the vicinity of Czorsztyn (west Carpathians) and its relationship to the Pleistocene. Summary of *Pliocenska Flora okolic Czorsztyna*, pp. 179–230. Warsaw: Wydawnictwa Geologiczne.

Tansley, A.G. (ed.) (1911). *Types of British vegetation*. Cambridge University Press.

Tansley, A.G. (1939). *The British Islands and their vegetation*. Cambridge University Press.

Tedrow, J.F.C. (1973). Soils of the polar region of North America. *Biuletyn Periglacjalny*, **23**, 157–165.

Thompson, K., Bakker, J. & Bekker, R. (1997). *The soil seed banks of North West Europe: methodology, density and longevity*. Cambridge University Press.

Traverse, A. (1994). Sedimentation of palynomorphs and palynodebris. In *Sedimentation of organic particles*, ed. A. Traverse, pp. 1–8. Cambridge University Press.

Turesson, G. (1925). The plant species in relation to habitat and climate. *Hereditas*, **6**, 147–236.

Turner, C. (1970). The Middle Pleistocene deposits at Marks Tey, Essex. *Philosophical Transactions of the Royal Society of London, B*, **257**, 373–440.

Turner, J., Hewetson, V.P., Hibbert, F.A., Lowry, K.H. & Chambers, C. (1973). The history of the vegetation and flora of Widdybank Fell and the Cow Green reservoir basin, Upper Teesdale. *Philosophical Transactions of the Royal Society of London, B*, **265**, 327–408.

Tutin, T.G. *et al.* (1964–1980). *Flora Europaea*, Vols. 1–5. Cambridge University Press.

Ukraintseva, V.V. (1986). On the composition of the forage of the large herbivorous mammals of the Mammoth Epoch: significance for palaeobiological and palaeogeographical reconstruction. *Quartärpaläeontologie*, **6**, 231–238.

Van der Hammen, T. (1951). Late-glacial flora and periglacial phenomena in the Netherlands. *Leidse Geologische Mededelingen*, **17**, 71–183.

References

Vandenberghe, J. (1992a). Periglacial phenomena and Pleistocene environmental conditions in the Netherlands – an overview. *Permafrost and Periglacial Processes*, **3**, 363–374.

Vandenberghe, J. (1992b). Geomorphology and climate of the cool oxygen isotope stage 3 in comparison with the cold stages 2 and 4 in The Netherlands. *Zeitschrift für Geomorphologie*, Suppl. **86**, 65–75.

Vandenberghe, J. (1993). Permafrost changes in Europe during the Last Glacial. *Permafrost and Periglacial Processes*, **4**, 121–135.

Van Huissteden, J. (1990). Tundra rivers of the Last Glacial: sedimentation and geomorphological processes during the Middle Pleniglacial in Twente, eastern Netherlands. *Mededelingen Rijks Geologische Dienst*, **44**, 1–138.

Vartanyan, S.L., Garutt, V.E. & Sher, A.V. (1993). Holocene dwarf mammoths from Wrangel Island in the Siberian Arctic. *Nature*, **362**, 337–340.

Vavrek, M.C., McGraw, J.B. & Yang, H.S. (1997). Within-population variation in demography of *Taraxacum officinale*: season- and size-dependent survival, growth and reproduction. *Journal of Ecology*, **85**, 277–287.

Walas, J. (1938). Wanderung der Gebirgspflanzen längs der Tatra-Flüsse. *Bulletin de l'Académie polonaise des Sciences et Lettres, Série B, Sciences Naturelles*, (1), 59–80.

Walker, D.A. & Everett, K.R. (1991). Loess ecosystems of northern Alaska: regional gradient and toposequence at Prudhoe Bay. *Ecological Monographs*, **61**, 437–464.

Walker, M.D., Walker, D.A., Everett, K.R. & Short, S.K. (1991). Steppe vegetation of south-facing slopes of pingos, central Arctic Coastal Plain, Alaska, U.S.A. *Arctic and Alpine Research*, **23**, 170–188.

Walter, H. (1985). *Vegetation of the Earth*. 3rd Ed. Berlin: Springer-Verlag.

Watt, A.S. (1964). The community and the individual. *Journal of Ecology*, **52** (Supplement), 203–211.

Watt, A.S., Perrin, R.M.S. & West, R.G. (1966). Patterned ground in Breckland: structure and composition. *Journal of Ecology*, **54**, 239–258.

Webb, D.A. (1978). Flora Europaea – a retrospect. *Taxon*, **27**, 3–14.

Webb, T. (1988). Climate and vegetation (review of Woodward, 1987). *Ecology*, **69**, 294–295.

Wesenberg-Lund, C. (1909). Om limnologiens betydning for Kwartærgeologien, særlig med hensyn til postglaciale tidsbestemmelser og temperaturangivelser. *Geologiska Föreningen i Stockholms Förhandlingar*, **31**.

West, R.G. (1980a). *The preglacial Pleistocene of the Norfolk and Suffolk coasts.* Cambridge University Press.

West, R.G. (1980b). Pleistocene forest history in East Anglia. *New Phytologist*, **85**, 571–622.

West, R.G. (1993). Devensian thermal contraction networks and cracks at Somersham, Cambridgeshire, UK. *Permafrost and Periglacial Processes*, **4**, 277–300.

West, R.G., Andrew, R., Catt, J.A., Hart, C.P., Hollin, J.T., Knudsen, K.-L., Miller, G.F., Penney, D.N., Pettit, M.E., Preece, R.C., Switsur, V.R., Whiteman, C.A. & Zhou, L.P. (1999). Late and Middle Pleistocene deposits at Somersham, Cambridgeshire, UK.: a model for reconstructing fluviatile/estuarine environments. *Quaternary Science Reviews*, **18**, 1247–1314.

West, R.G., Andrew, R. & Pettit, M.E. (1993). Taphonomy of plant remains on floodplains of tundra rivers, present and Pleistocene. *New Phytologist*, **123**, 203–221.

West, R.G., Funnell, B.M. & Norton, P.E.P. (1980). An Early Pleistocene cold marine episode in the North Sea: pollen and faunal assemblages at Covehithe, Suffolk, England. *Boreas*, **9**, 1–10.

West, R.G., Dickson, C.A., Catt, J.A., Weir, A.H. & Sparks, B.W. (1974). Late Pleistocene deposits at Wretton, Norfolk. II. Devensian deposits. *Philosophical Transactions of the Royal Society of London, B*, **267**, 337–420.

Whittaker, A., Cope, J.C.W., Cowie, J.W., Gibbons, W., Hailwood, E.A., House, M.R., Jenkins, D.G., Rawson, P.F., Rushton, A.W.A., Smith, D.G., Thomas, A.T. & Wimbledon, W.A. (1991). A guide to stratigraphic procedure. *Journal of the Geological Society*, **148**, 813–824.

Williams, R.B.G. (1975). The British climate during the last glaciation: an interpretation based on periglacial phenomena. In *Ice ages ancient and modern*, ed. A.E. Wright & F. Moseley, pp. 95–120. Liverpool: Seel House.

Williamson, M. (1996). *Biological invasions.* London: Chapman & Hall.

Wilmott, A.J. (1935). Evidence in favour of survival of the British Flora in glacial times. *Proceedings of the Royal Society of London, B*, **118**, 215–222.

Woillard, G. (1975). Recherches palynologiques sur le Pleistocene dans l'est de la Belgique et dans le Vosges Lorraine. *Acta Geographica Lovaniensia*, **14**, 1–118.

Woillard, G.M. (1978). Grande Pile Peat Bog: a continuous pollen record for the last 140,000 years. *Quaternary Research*, **9**, 1–21.

Wolfe, J.A. (1985). Distribution of major vegetational types during the Tertiary. In *The carbon cycle and atmospheric CO_2: natural variations Archaean to present. Geophysical Monograph*, **32**, 357–375.

Woodland, A.W. (1946). Water supply from undergound sources of Cambridge-Ipswich District. Part X. General Discussion. *Geological Survey of Great Britain, Wartime Pamphlet* No. 20.

Woodward, F.I. (1987). *Climate and plant distribution.* Cambridge University Press.

Young, S.B. (1971). The vascular flora of Saint Lawrence Island, with special reference to floristic zonation in the Arctic regions. *Contributions from the Gray Herbarium of Harvard University*, No. **201**, 11–115.

Young, S.B. (1982). The vegetation of land-bridge Beringia. In *Palaeoecology of Beringia*, ed. D.M. Hopkins *et al.*, pp. 179–191. London: Academic Press.

Young, S.B. (1989). *To the Arctic.* New York: Wiley.

Yurtsev, B.A. (1982). Relics of the xerophyte vegetation of Beringia in northeastern Asia. In *Palaeoecology of Beringia*, ed. D.M. Hopkins *et al.*, pp. 157–177. London: Academic Press.

Yurtsev, B.A. (1985). Beringia and its biota in the Late Cenozoic: a synthesis. In *Beringia in the Cenozoic Era*, ed. V.L. Kontrimavichus, pp. 261–275. Rotterdam: Balkema.

Zagwijn, W.H. (1961). Vegetation, climate and radiocarbon datings in the Late Pleistocene of The Netherlands. Part 1. Eemian and Early Weichselian. *Mededelingen van de Geologischen Stichting*, n.s. **14**, 15–45.

Zagwijn, W.H. (1975). Indeling van het Kwartair op grond van veranderingen in vegetatie en klimaat. In *Toelichting bij geologische overzichtskaarten van Nederland*, ed. W.H. Zagwijn & C.J. van Staalduinen, pp. 109–114. Haarlem: Rijks Geologische Dienst.

Zimov, S.A., Chuprynin, V.I., Oreshko, A.P., Chapin, F.S., Reynolds, J.F. & Chapin, M.C. (1995). Steppe-tundra transition: a herbivore-driven biome shift at the end of the Pleistocene. *American Naturalist*, **146**, 765–794.

Appendix I Works consulted in the identification of macroscopic remains

Beijerinck, W. 1947. *Zadenatlas Der Nederland Flora.* Wageningen: Veenman & Zonen.

Berggren, G. 1969. *Atlas of seeds and small fruits of Northwest-European plant species, Part 2, Cyperaceae.* Stockholm: Swedish Museum of Natural History.

Berggren, G. 1981. *Atlas of seeds and small fruits of Northwest-European plant species, with morphological descriptions, Part 3, Salicaceae – Cruciferae.* Stockholm: Swedish Museum of Natural History.

Clapham, A.R., Tutin, T.G. & Warburg, E.F. 1969. *Flora of the British Isles.* 2nd Ed. Cambridge University Press.

Clapham, A.R., Tutin, T.G. & Moore, D.M. 1989. *Flora of the British Isles.* 3rd Ed. Cambridge University Press.

Godwin, H. 1975. *History of the British Flora – a factual basis for phytogeography.* 2nd Ed. Cambridge University Press.

Hultén, E. 1971. *Atlas of the distribution of vascular plants in northwestern Europe.* Stockholm: AB Kartografiska Institutet.

Jermy, A.C., Chater, A.O. & David, R.W. 1982. *Sedges of the British Isles.* Handbook No. 1, 2nd Ed. London: Botanical Society of the British Isles.

Katz, N.Ja., Katz, S.V. & Kipiani, M.G. 1965. *Atlas and keys of fruits and seeds occurring in the Quaternary deposits.* Moscow: Nauka.

Körber-Grohne, U. 1964. *Bestimmungsschlüssel für subfossile Juncus-Samen und Gramineen-Früchte.* Hildesheim: August Lax.

Meikle, R.D. 1984. *Willows and Poplars of Great Britain and Ireland.* Handbook No. 4. London: Botanical Society of the British Isles.

Nordhagen, R. 1970. *Norsk Flora.* Oslo: H. Aschehoug and Co.

Perring, F.H. & Walters, S.M. (eds.) 1990. *Atlas of the British Flora.* 3rd Ed. London: Botanical Society of the British Isles.

Polunin, N. 1959. *Circumpolar Arctic Flora.* Oxford: Clarendon Press.

Rich, T.C.G. 1991. *Crucifers of Great Britain and Ireland.* Handbook No. 6. London: Botanical Society of the British Isles.

Ross-Craig, S. 1948–1973. *Drawings of British Plants, parts 1-31.* London: G. Bell and Sons.

Tutin, T.G. 1980. *Umbellifers of the British Isles.* Handbook No. 2. London: Botanical Society of the British Isles.

Tutin, T.G. *et al.* 1964–1980. *Flora Europaea,* Vols. 1-5. Cambridge University Press.

Appendix II Taxa recorded in cold stage sediments, in *Flora Europaea* order (abbreviations as in Table 4.3, p. 43)

PTERIDOPHYTA
Pteridophyta   m

1. LYCOPODIACEAE
   Lycopodiaceae   p
   Lycopodium   p
   Lycopodium cf.   p

   Diphasium   p

      * *Huperzia selago*   p
      * *L. annotinum*   p
      L. cf. *annotinum*   p
      cf. *L. annotinum*   p
      * *Diphasium alpinum*   p
      D. cf. *alpinum*   p

2. SELAGINELLACEAE
   Selaginella   p

      * *S. selaginoides*   m,p
      cf. *S. selaginoides*   m
      S. type z   m

3. ISOETACEAE
   Isoetes   m
   cf. *Isoetes*   m

      * *I. lacustris*   m
      I. cf. *histrix*   m

4. EQUISETACEAE
   Equisetum   m,p
   cf. *Equisetum*   m

5. OPHIOGLOSSACEAE
   Ophioglossum   p
   Botrychium   p

6. OSMUNDACEAE
   Osmunda   m,p

      O. claytoniana t nfe   p
      * *O. regalis*   p

7–22. POLYPODIACEAE
   Filicales (Polypodiaceae s.l.)   m,p

285

Taxa recorded in cold stage sediments, in Flora Europaea *order* (*cont.*)

10. CRYPTOGRAMMACEAE
   Cryptogramma   p

13. HYPOLEPIDACEAE
   Pteridium   p   * *P. aquilinum*   m,p

18. ATHYRIACEAE
                                               * *Cystopteris fragilis*   p
                                               C. fragilis t   p
                                               cf. *C. fragilis* t   p

19. ASPIDIACEAE
   Dryopteris t   m   *D. filix-mas* t   p
   cf. *Polystichum* t   p   * *Gymnocarpium dryopteris*   p

22. POLYPODIACEAE
   Polypodiaceae   p
   Polypodium   p   * *P. vulgare*   p

24. SALVINIACEAE
                                               Salvinia natans   m

25. AZOLLACEAE
   Azolla   m,p   *A. filiculoides*   m
                                               A. cf. *filiculoides*   m

SPERMATOPHYTA

26–30. CONIFERAE
   Coniferae   m
   Coniferae ?   m

26. PINACEAE
   Pinaceae   m
   Abies   p   * *A. alba*   m
   Abies sp.   m   *A.* cf. *alba*   m
   Tsuga   p
   cf. *Tsuga*   p
   Picea   m,p   * *P. abies*   m
   cf. *Picea*   m   ** *P. abies* ssp. *obovata*   m
                                               P. abies cf. *obovata*   m
   Pinus   m,p   * *P. sylvestris*   m
   Pinus (sacs)   p   *P.* cf. *sylvestris*   m
   cf. *Pinus*   m   cf. *P. sylvestris*   m
                                               P. sylvestris t   p
   Pinus subg. *Haploxylon*   p
   Pinus subg. *Pinus*   m

27. TAXODIACEAE
   Taxodiaceae   p

28. CUPRESSACEAE
   Cupressaceae p
   Cupressaceae b   p
      Juniperus m,p   * *J. communis*   m,p
      cf. *Juniperus*   m,p

29. TAXACEAE
      Taxus p   *T. baccata*   m

30. EPHEDRACEAE
      Ephedra p   *E. distachya* t   p

DICOTYLEDONES

31. SALICACEAE
   Salicaceae m
      Salix m,p   * *S. reticulata*   m
      cf. *Salix*   m,p   *S.* cf. *reticulata*   m
      Salix spp.   m   * *S. herbacea*   m,p
      cf. *Salix* spp.   m   cf. *S. herbacea*   m
      Salix dwarf spp.   m   *S.* cf. *herbacea*   m
      Salix spp. incl. *S. polaris* hybrid   m   *S. herbacea* t   p
      Salix incl. *S. herbacea*   p   *S.* cf. *herbacea* × *repens*   m
                                          * *S. polaris*   m
                                          S. cf. *polaris*   m
                                          * *S. myrsinites*   m
                                          S. cf. *myrsinites*   m
                                          * *S. lanata*   m
                                          * *S. phylicifolia*   m
                                          S. cf. *phylicifolia*   m
                                          * *S. repens*   m
                                          S. cf. *repens*   m
                                          * *S. arbuscula*   m
                                          S. cf. *arbuscula*   m
                                          * *S. lapponum*   m
                                          S. cf. *lapponum*   m
                                          * *S. viminalis*   m,p
                                          S. cf. *viminalis*   m
                                          S. viminalis t   p
      Populus p   *Populus* cf. *tremula*   m
      cf. *Populus*   m

32. MYRICACEAE
      Myrica p

33. JUGLANDACEAE
      Juglans p
      Carya p
      Pterocarya p

34. BETULACEAE
      Betula m,p   * *B. pendula*   m
      cf. *Betula*   m   *B.* cf. *pendula*   m

Taxa recorded in cold stage sediments, in Flora Europaea *order* (*cont.*)

Betula spp. m	*B. pendula* or *pubescens* m
	cf. *B.* tree spp. × *B. nana* m
	* *B. pubescens* m
	B. cf. *pubescens* m
	* *B. nana* m,p
	B. cf. *nana* m
	cf. *B. nana* m
	B. nana t p
Alnus m,p	* *A. glutinosa* m
cf. *Alnus* m	

35. CORYLACEAE
   Carpinus p   *C. betulus*   m
   Ostrya t   p
   Corylus p   *C. avellana*   m
   Coryloid p

36. FAGACEAE
   Fagus p
   Quercus   m,p

37. ULMACEAE
   Ulmus   p

40. URTICACEAE
   Urtica   p   * *Urtica dioica*   m
                                Urtica dioica ?   m

47. POLYGONACEAE
   Polygonaceae p
                                * *Koenigia islandica*   p
   Polygonum   m,p   * *P. maritimum*   p
   cf. *Polygonum*   m   + *P. oxyspermum* agg.   m
                                * *P. aviculare*   m,p
                                + *P. aviculare* agg.   m,p
                                P. cf. *aviculare* agg.   m
                                P. cf. *aviculare*   m
                                P. aviculare t   p
                                * *P. hydropiper*   m
                                P. hydropiper ?   m
                                * *P. persicaria*   p
                                P. persicaria group   p
                                P. persicaria t   p
                                * *P. lapathifolium*   m
                                * *P. amphibium*   m
                                P. amphibium t   p
                                * *P. bistorta*   p
                                P. bistorta t   p
                                P. bistorta/viviparum t   p
                                * *P. viviparum*   m,p

* *Bilderdykia convolvulus* m,p
B. *convolvulus* t   p
* *Oxyria digyna* m

Rumex m,p
Rumex ssp. m
cf. *Rumex* m
Rumex subg. *Rumex* m,p

* *R. angiocarpus* m
* *R. tenuifolius* m
* *R. acetosella* m,p
R. cf. *acetosella* m
+ *R. acetosella* agg. m
R. cf. *acetosella* agg. m
R. *acetosella* t   m,p
* *R. acetosa* m,p
R. *acetosa* t   p
R. *hydrolapathum/aquaticus/
longifolius* m
* *R. crispus* p
R. cf. *crispus* m
R. *crispus* ? m
R. *crispus* t   p
* *R. conglomeratus* m
* *R. palustris* m
* *R. maritimus* m
R. cf. *maritimus* m

48. CHENOPODIACEAE
    Chenopodiaceae m,p
    Chenopodium m
    Chenopodium spp. m
    cf. *Chenopodium* m
    Chenopodium sect.
        Pseudoblitum CTW   m
    Chenopodium cf. sect.
        Pseudoblitum CTW   m
    Atriplex m
    cf. *Atriplex* m
    Atriplex spp. m
    cf. *Atriplex* spp. m

* *C. rubrum* m
# *C. botryodes* m
* *C. murale* m
* *C. ficifolium* m
* *C. album* m
C. cf. *album* m
C. *album* t   m
* *A. patula* m
A. cf. *patula* m
* *A. hastata* m
A. *hastata* t   m
A. cf. *hastata* m
* *A. glabriuscula* m
C. cf. *hyssopifolium* m,p

    Corispermum m
    cf. *Corispermum* m
    Salicornia m
    cf. *Salicornia* m

* *Suaeda maritima* m

55. PORTULACACEAE

* *Montia fontana* m,p
** *Montia fontana* ssp. *fontana* m
+ *Montia fontana* agg. m

57. CARYOPHYLLACEAE
    Caryophyllaceae m,p
    cf. Caryophyllaceae m

Taxa recorded in cold stage sediments, in Flora Europaea *order* (*cont.*)

Arenaria m	# *A. norvegica* m
Arenaria spp. m	** *A. norvegica* ssp. *norvegica* m
cf. *Arenaria* m	* *A. ciliata* m
	+ *A. ciliata* agg. B m
	A. cf. *ciliata* m
	cf. *A. ciliata* m
	* *A. gothica* m
	* *A. serpyllifolia* m
	** *A. serpyllifolia* ssp. *macrocarpa* m
	A. cf. *serpyllifolia* m
	* *Moehringia trinervia* m
	M. trinervia g p
	M. trinervia t p
Minuartia m	*M. hybrida* g p
	* *M. verna* m
	cf. *M. verna* m
	* *M. rubella* m
	M. cf. *rubella* m
	cf. *M. rubella* m
	* *M. stricta* m
	M. cf. *stricta* m
	* *M. sedoides* m
Stellaria m	*S*. cf. *nemorum* m
cf. *Stellaria* m	* *S. media* m
	S. cf. *media* m
	cf. *S. media* m
	* *S. holostea* p
	S. cf. *alsine* m
	* *S. palustris* m
	S. cf. *palustris* m
	* *S. graminea* m
	S. cf. *graminea* m
	cf. *S. graminea* m
	* *S. crassifolia* m
	S. cf. *crassifolia* m
	cf. *S. crassifolia* m
	* *Holosteum umbellatum* m
Cerastium m	*C. cerastoides* g p
Cerastium t p	* *C. arvense* m
cf. *Cerastium* m	*C*. cf. *arvense* m
	cf. *C. arvense* m
	C. arvense g p
	* *C. alpinum* m
	C. alpinum s.l. m
	** *C. fontanum* m
	C. fontanum t p
	** *C. fontanum* ssp. *triviale* m
	* *Myosoton aquaticum* m
	* *Sagina cespitosa* m

	S. cf. *procumbens* m
	S. *procumbens* t p
Scleranthus m,p	* S. *perennis* p
Scleranthus spp. m	* S. *annuus* m
	S. cf. *annuus* m
Herniaria m	* H. *glabra* m
	cf. *Spergularia medialmarina* m
Lychnis m	* L. *flos-cuculi* m
	* L. *alpina* m
	cf. L. *alpina* m
	* L. *triflora* NP m
Silene m	S. *paradoxa* ? m
Silene t p	S. cf. *wahlbergella* m
cf. *Silene* m	S. cf. *wahlbergellalfurcata* p
	* S. *furcata* m
	* S. *vulgaris* m,p
	+ S. *vulgaris* agg. m
	S. *vulgaris* ? m
	** S. *vulgaris* ssp. *maritima* m
	S. *vulgaris* cf. ssp. *maritima* m
	S. *vulgaris* t p
	* S. *acaulis* m
	S. *acaulis* ? m
	cf. S. *acaulis* m
	* S. *dioica* m
	S. *dioica* t p
Dianthus spp. m	D. cf. *gratianopolitanus* m
	* D. *deltoides* m
	cf. D. *deltoides* m
	D. cf. *carthusianorum* m

58. NYMPHAEACEAE
    Nymphaea   p   * N. *alba*   m,p
    Nuphar   m,p   * N. *lutea*   m
    cf. *Nuphar*   p

60. CERATOPHYLLACEAE
    Ceratophyllum   m,p   * C. *demersum*   m
                       C. cf. *submersum*   m

61. RANUNCULACEAE
  Ranunculaceae   p
    Caltha   p   * C. *palustris*   m,p
    Caltha t
    Ranunculus   m,p   * R. *nemorosus*   m
    Ranunculus spp.   m,p   * R. *repens*   m
    Ranunculus t   p   R. cf. *repens*   m
    cf. *Ranunculus*   m   R. *repens* t   p
    Ranunculus subg. *Ranunculus*   m * R. *acris*   m,p
    cf. *Ranunculus* subg.   R. *acris* t   m,p
      Ranunculus   m   R. *acris* g   p

Taxa recorded in cold stage sediments, in Flora Europaea *order (cont.)*

Ranunculus sect. *Auricomus* m	*R.* cf. *acris* m
Ranunculus cf. sect.	* *R. bulbosus* m
Auricomus m	* *R. sardous* m
Ranunculus sect.	*R.* cf. *sardous* m
Aconitifolii m	* *R. parviflorus* m
Ranunculus cf. sect.	*R.* cf. *nivalis/pygmaeus* m
Aconitifolii m	* *R. hyperboreus* m
Ranunculus sect.	*R.* cf. *hyperboreus* m
Chrysantha CTW m	* *R. sceleratus* m
Ranunculus Batrachium m,p	* *R. aconitifolius* m
Ranunculus Batrachium t p	*R. aconitifolius*? m
	* *R. platanifolius* m
	R. glacialis g p
	* *R. flammula* m
	R. cf. *flammula* m
	R. flammula t m
	* *R. reptans* m
	R. cf. *reptans* m
	* *R. lingua* m
	R. cf. *lingua* m
	* *R. hederaceus* m
	* *R. aquatilis* m
	R. cf. *aquatilis* p
	R. trichophyllus t p
	* *Myosurus minimus* m
Thalictrum m,p	* *T. alpinum* m
cf. *Thalictrum* m	*T.* cf. *alpinum* m
	cf. *T. alpinum* m
	T. alpinum g p
	* *T. minus* m
	T. cf. *minus* m
	+ *T. minus* agg. m
	* *T. flavum* m
	T. flavum g p
67. PAPAVERACEAE	
cf. *Papaver* m	
Papaver sect. *Scapiflora* m	+ *P. radicatum* s.l. m,p
68. CRUCIFERAE	
Cruciferae m,p	
cf. Cruciferae m	
	cf. *Descurainia sophia* m
	* *Erysimum cheiranthoides* m
Barbarea m	# *B. vulgaris* m
	B. cf. *vulgaris* m
	# *B. stricta* m
Rorippa m	* *R. sylvestris* m
cf. *Rorippa* m	cf. *Rorippa sylvestris* m

cf. *R. sylvestris* or *R. amphibia*   m
* *R. islandica*   m
cf. *R. islandica*   m
* *Nasturtium microphyllum*   m
N. cf. *microphyllum*   m

Cardamine   m   *C.* cf. *amara*
Cardamine spp.   m   * *C. pratensis*   m
cf. *Cardamine*   m   *C.* cf. *pratensis*   m
cf. *Cardamine* nb sp.   m   cf. *C. pratensis*   m,p
                             cf. *Cardaminopsis petraea*   m
Arabis   m   # *A. hirsuta*   m
Arabis type *A*   m   *A.* cf. *hirsuta*   m
cf. *Arabis*   m   # *A. stricta*   m
cf. *Arabis* spp.   m   * *A. alpina*   m
cf. *Arabis* t   m
Alyssum   m   *A. (? saxatile)*   m
Alyssum nb sp.   m
cf. *Alyssum* nb sp.   m
Alyssum t   m
cf. *Alyssum*   m
Draba   m   *D.* nb cf. *alpina* group   m
Draba spp.   m   * *D. norwegica*   m
cf. *Draba*   m   *D.* cf. *incana/norwegica*   m
                             D. incana/norwegica t   m
                             * *D. incana*   m
                             D. cf. *incana*   m
                             D. incana t   m
                             cf. *D. incana*   m
Erophila   m   * *E. verna*   m
                             + *E. verna* agg.   m
                             cf. *E. verna*   m
                             ** *E. verna* ssp. *spathulata*   m
                             E. verna cf. ssp. *spathulata*   m
                             cf. *E. verna* ssp. *spathulata*   m
Cochlearia   m   # *C. danica*   m
                             + *C. officinalis* agg.   m
                             C. cf. *officinalis*   m
                             * *C. pyrenaica*   m
                             C. cf. *pyrenaica*   m
                             C. pyrenaica ?   m
                             * *Capsella bursa-pastoris*   m
                             C. bursa-pastoris ?   m
Hornungia t   p
                             * *Thlaspi arvense*   m
                             * *Coronopus squamatus*   m
Diplotaxis   m   * *D. tenuifolia*   m
cf. *Diplotaxis*   m   cf. *D. tenuifolia*   m
Sinapis t   p

71. DROSERACEAE
    Drosera   p

Taxa recorded in cold stage sediments, in Flora Europaea *order (cont.)*

72. CRASSULACEAE
    Sedum t p
    cf. *Sedum* p
                  * *Rhodiola rosea* m

73. SAXIFRAGACEAE
    Saxifraga m,p   *S.* cf. *nivalis* p
    Saxifraga spp. m   *S. stellaris* t p
    cf. *Saxifraga* m   *S.* cf. *stellaris* m
    Saxifraga sect.   *S.* cf. *hirsuta* m
        Dactyloides CTW m   *S. hirsuta* t p
    Saxifraga cf. sect.   * *S. hirculus* m
        Sedoides CTW m   * *S. tridactylites* m
                           S. sedoides t m
                          * *S. cespitosa* m
                          * *S. rosacea* m
                           S. cf. *rosacea* m
                           S. hypnoides/rosacea m
                        + *S. hypnoides* agg. m
                           S. hypnoides type FGB m
                           S. cf. *hypnoides* m
                           S. granulata t p
                        * *S. oppositifolia* m,p
                           S. cf. *oppositifolia* m
                           S. oppositifolia t p
                           Chrysosplenium alterniflorum t p

74. PARNASSIACEAE
    Parnassia p   * *P. palustris* m,p
                           cf. *P. palustris* m

79. PLATANACEAE
    Platanus p

80. ROSACEAE
    Rosaceae p
    cf. Rosaceae m
    Filipendula m,p   * *F. ulmaria* m,p
    Rubus m   * *R. chamaemorus* p
    Rubus spp. m   * *R. idaeus* m
    cf. *Rubus* m   * *R. fruticosus* CTW m
    Rosa t p
    Sanguisorba? m   * *S. officinalis* m,p
                           cf. *S. officinalis* m,p
                        ** *S. minor* ssp. *minor* m,p
    Dryas p   * *D. octopetala* m
    Geum p
    cf. *Geum* m
    Potentilla m,p   * *P. fruticosa* m

Potentilla spp.   m
cf. *Potentilla*   m

P. cf. *fruticosa*   m
cf. *P. fruticosa*   m
* *P. palustris*   m
P. cf. *palustris*   m
* *P. anserina*   m
# *P. nivea*   m
* *P. argentea*   m
P. cf. *argentea*   m
* *P. crantzii*   m
P. cf. *crantzii*   m
P. crantzii t   m
cf. *P. crantzii* t   m
cf. *P. crantzii*   m
* *P. tabernaemontani*   m
P. cf. *tabernaemontani*   m
* *P. erecta*   m
P. erecta t   m
* *P. reptans*   m
* *P. sterilis*   m
# *Sibbaldia procumbens*   m
cf. *Fragaria vesca*   m

Alchemilla   m,p
cf. *Aphanes*   m

* *A. arvensis*   m
* *A. microcarpa*   m

Sorbus   p

Prunus cf. *avium*   m

81. LEGUMINOSAE
Leguminosae   m,p
cf. *Leguminosae*   m

cf. *Genista anglica*   m

cf. *Ulex* t   p

Astragalus danicus t   p
* *A. alpinus*   p

cf. *Oxytropis*   p
Vicia   m,p
Vicia ?   m
Lathyrus   m
Lathyrus ?   m
Lathyrus/Vicia t   p
Ononis t   p
Medicago   m
cf. *Medicago*   m

V. cf. *cracca*   m
* *V. sylvatica*   m
* *L. sylvestris*   m

* *M. lupulina*   m
** *M. sativa* ssp. *falcata*   m
M. arabica/minima   m
T. cf. *campestre*   m
T. cf. *pratense*   p

Trifolium   m,p
cf. *Trifolium*   m
Trifolium t   p
Lotus   p
Lotus t   p

L. uliginosus t   p
L. cf. *uliginosus*   m
* *Anthyllis vulneraria*   p

Onobrychis t   p

* *O. viciifolia*   m

Taxa recorded in cold stage sediments, in Flora Europaea *order (cont.)*

83. GERANIACEAE
  cf. Geraniaceae   m
  Geranium   p
  Geranium spp.   m
  cf. *Geranium*   m
  Geranium/Erodium   p

86. LINACEAE
  cf. *Linum*   m   * *L. perenne*   m,p
                                          L. cf. *perenne*   m
                                          ** *L. perenne* ssp. *anglicum*   m,p
                                          + *L. perenne* agg.   m
                                          L. cf. *perenne* agg.   m
                                          cf. *L. perenne*   p
                                          L. austriacum t   p
                                          * *L. catharticum*   m,p
                                          cf. *L. catharticum*   p

87. EUPHORBIACEAE
                                          * *Mercurialis perennis*   m
  Euphorbia   m,p   * *E. cyparissias*   m
                                          E. cf. *cyparissias*   m

95. ACERACEAE
  Acer   p

99. AQUIFOLIACEAE
  Ilex   p

102. BUXACEAE
  Buxus   p

103. RHAMNACEAE
  Rhamnus   p
  Frangula   p   * *F. alnus*   m

105. TILIACEAE
  Tilia   p   *T. cordata*   p

108. ELAEAGNACEAE
  Hippophae   p   * *H. rhamnoides*   m,p

109. GUTTIFERAE
  Hypericum   p   * *H. tetrapterum*   m

110. VIOLACEAE
  Viola   m   # *V. odorata*   m
  Viola spp.   m   # *V. hirta*   m
  Viola subg. *Melanium*   m   # *V. reichenbachiana*   m

cf. *Viola*   m,p

\# *V. riviniana*   m
V. cf. *riviniana*   m
V. riviniana t   m
\# *V. canina*   m
V. cf. *canina*   m
V. canina ?   m
* *V. palustris*   m
cf. *V. palustris*   p
* *V. lutea*   m
* *V. tricolor*   m
V. lutea/tricolor t   m
V. cf. *arvensis*   m

112. CISTACEAE
   Helianthemum   m,p
   cf. *Helianthemum*   m

* *H. canum*   m
H. cf. *canum*   m

115. ELATINACEAE
   Elatine   m

* *E. hydropiper*   m
* *E. hexandra*   m

119. LYTHRACEAE

* *Lythrum salicaria*   p

120. TRAPACEAE
   cf. *Trapa*   m

123. ONAGRACEAE
   Epilobium   m,p
   Epilobium t   p

* *E. parviflorum*   m
* *E. alsinifolium*   m

124. HALORAGACEAE
   Myriophyllum   m,p

* *M. verticillatum*   m,p
* *M. spicatum*   m,p
M. cf. *spicatum*   m
* *M. alterniflorum*   m,p
cf. *M. alterniflorum*   m

126. HIPPURIDACEAE
   Hippuris   m,p

* *H. vulgaris*   m
cf. *H. vulgaris*   m

127. CORNACEAE

Cornus sanguinea   m
cf. *C. sanguinea*   m
C. mas t   p
* *C. suecica*   p
cf. *C. suecica*   m

128. ARALIACEAE
   Hedera   p

Taxa recorded in cold stage sediments, in Flora Europaea *order* (*cont.*)

129. UMBELLIFERAE
    Umbelliferae m,p
      Hydrocotyle p
                    * *H. vulgaris* m
                    cf. *Chaerophyllum temulentum* p
                    * *Anthriscus sylvestris* m
                    A. sylvestris t p
                    cf. *Bunium bulbocastanum* m
                    * *Pimpinella saxifraga* m
      Sium t p
      Berula t p
                    * *B. erecta* m
                    Seseli libanotis t p
      cf. *Oenanthe* m
                    * *O. aquatica* m
      Oenanthe ?
                    Meum athamanticum t p
                    Bupleurum cf. *falcatum* m
      Apium/*Berula* t p
                    * *Apium nodiflorum* m
                    * *A. inundatum* m
                    A. inundatum t p
                    * *Cicuta virosa* m
                    cf. *Angelica sylvestris* m
      Peucedanum ? m
                    P. palustre g p
      Pastinaca p
                    * *P. sativa* m
      Heracleum p
                    * *H. sphondylium* m,p
      Heracleum/*Pastinaca* p
                    * *Daucus carota* p

132/3. ERICALES
    Ericales p

132. ERICACEAE
    Ericaceae m,p
    cf. Ericaceae m
      Erica m
                    E. cf. *tetralix* m
                    cf. *E. tetralix* m
      Bruckenthalia p
                    * *B. spiculifolia* m
      cf. *Bruckenthalia* m
      Calluna p
                    * *C. vulgaris* m,p
      cf. *Calluna* m
      Rhododendron p
                    * *R. ponticum* m
                    cf. *Loiseleuria procumbens* t p
                    * *Arctostaphylos uva-ursae* m
      Vaccinium m
                    * *V. oxycoccos* m
      cf. *Vaccinium* m
                    V. cf. *oxycoccos* m
                    cf. *V. oxycoccos* m
                    * *V. myrtillus* m
                    V. cf. *myrtillus* m

133. EMPETRACEAE
      Empetrum m,p
                    * *E. nigrum* m

cf. *Empetrum*   p

** *E. nigrum* ssp. *nigrum*   m
+ *E. nigrum* agg.   m
E. nigrum t   p

135. PRIMULACEAE
 cf. Primulaceae   m
 Primula   m
 cf. *Primula*   m
 Primula subg. *Aleuritia*   m
 Primula sect. *Aleuritia* CTW   m

\# *P. elatior*   m
\# *P. veris*   m
P. farinosa ?   m
P. cf. *farinosa*   m
* *P. scotica*   m
* *Androsace septentrionalis*   m
* *Lysimachia vulgaris*   m
L. vulgaris t   p
* *L. thyrsiflora*   m
* *Glaux maritima*   m
cf. *G. maritima*   m

136. PLUMBAGINACEAE
 Armeria   m,p
 Armeria t   p
 cf. *Armeria*   m

* *A. maritima*   m,p
+ *A. maritima* s.l.   m
A. cf. *maritima*   m
cf. *A. maritima*   m
cf. *Limonium vulgare*   m

139. OLEACEAE
 Fraxinus   p

140. GENTIANACEAE
 Gentianaceae   p

 Gentianella   m

* *Centaurium erythraea*   m
Gentiana cf. *purpurea*   p
* *G. pneumonanthe*   p
G. pneumonanthe t   p
G. cf. *pneumonanthe*   p
* *G. nivalis*   p

Lomatogonium rotatum t   p

141. MENYANTHACEAE
 Menyanthes   m,p

* *M. trifoliata*   m,p

144. RUBIACEAE
 Rubiaceae   p
 Galium   p
 cf. *Galium*   m

145. POLEMONIACEAE
 Polemonium   p
 cf. *Polemonium*   p

* *P. caeruleum*   p

146. CONVOLVULACEAE
 cf. *Cuscuta*   m
 Convolvulus   p

* *C. arvensis*   p

Taxa recorded in cold stage sediments, in Flora Europaea *order (cont.)*

148. BORAGINACEAE
　Boraginaceae　p
　　Lithospermum　p
　　　　　　　　　　　　　　* *Symphytum officinale*　p
　　Myosotis　m
　　cf. *Myosotis*　p

150. CALLITRICHACEAE
　　Callitriche　m　　　　　* *C. hermaphroditica*　m
　　cf. *Callitriche*　m　　　　*C.* cf. *stagnalis*　m
　　　　　　　　　　　　　* *C. obtusangula*　m
　　　　　　　　　　　　　* *C. platycarpa*　m
　　　　　　　　　　　　　C. cf. *platycarpa*　m
　　　　　　　　　　　　　C. cf. *palustris*　m

151. LABIATAE
　Labiatae　m,p
　　cf. *Ajuga*　p　　　　　* *Ajuga reptans*　m
　　　　　　　　　　　　　A. cf. *reptans*　m
　　cf. *Scutellaria*　p
　　Galeopsis　m　　　　　+ *G. tetrahit* agg.　m
　　　　　　　　　　　　　G. cf. *tetrahit* agg.　m
　　Stachys　m　　　　　* *S. sylvatica*　m
　　Stachys t　p　　　　　* *S. palustris*　m
　　Nepeta?　m
　　Prunella t　p　　　　　* *Prunella vulgaris*　m
　　　　　　　　　　　　　cf. *Acinos arvensis*　m
　　　　　　　　　　　　** *Calamintha sylvatica* ssp. *ascendens* m
　　cf. *Origanum*　p　　　　* *O. vulgare*　m
　　　　　　　　　　　　　* *Lycopus europaeus*　m
　　Mentha　m　　　　　# *M. arvensis*　m
　　Mentha spp.　m　　　　*M.* cf. *arvensis*　m
　　cf. *Mentha*　m　　　　* *M. aquatica*　m
　　Mentha t　p　　　　　*M.* cf. *aquatica*　m

152. SOLANACEAE
　　　　　　　　　　　　　* *S. dulcamara*　m

154. SCROPHULARIACEAE
　Scrophulariaceae　p
　cf. Scrophulariaceae　m
　　　　　　　　　　　　　Scrophularia cf. *nodosa*　p
　　　　　　　　　　　　　* *Linaria vulgaris*　m
　　Veronica　m　　　　　# *V. anagallis-aquatica*　m
　　cf. *Veronica*　m　　　　*V.* cf. *anagallis-aquatica*　m
　　　　　　　　　　　　　# *V. catenata*　m
　　　　　　　　　　　　　V. cf. *spicata*　m
　　Bartsia　m,p
　　　　　　　　　　　　　* *Pedicularis lanata*
　　　　　　　　　　　　　　Cham. & Schlect nfe　m

                * *P. hirsuta*   m
                * *P. palustris*   m

Rhinanthus   m
Rhinanthus t   p
cf. *Rhinanthus*   m

161. LENTIBULARIACEAE
    Utricularia   p

163. PLANTAGINACEAE
    Plantago   m,p   * *P. major*   m
    Plantago spp.   m   cf. *P. major*   p
                         P. major/media   p
                         P. major/media t   p
                * *P. coronopus*   p
                         P. cf. *coronopus*   m
                * *P. maritima*   m,p
                         P. cf. *maritima*   m
                         P. maritima t   p
                * *P. media*   m,p
                         P. cf. *media*   p
                * *P. lanceolata*   p
    Littorella   m,p   * *L. uniflora*   m

164. CAPRIFOLIACEAE
    cf. *Sambucus*   p   *S. nigra*   m
                         cf. *S. nigra*   m
                # *S. racemosa*   m
                         S. cf. *racemosa*
                * *Lonicera xylosteum*   p

166. VALERIANACEAE
                * *Valerianella dentata*   m
    Valeriana   m,p   * *Valeriana officinalis*   m,p
                         V. cf. *officinalis*   m
                * *V. dioica*   m,p
                         V. cf. *dioica*   m
                         cf. *V. dioica*   p

167. DIPSACACEAE
    Succisa   p   * *S. pratensis*   p
    Scabiosa   m,p   * *S. columbaria*   m,p
                         cf. *S. columbaria*   m

168. CAMPANULACEAE
    Campanulaceae   m,p
       Campanula   m,p   *C.* cf. *patula*   m
       Campanula t   p   * *C. glomerata*   m
       cf. *Campanula*   m   * *C. rotundifolia*   m
                         C. cf. *rotundifolia*   m
       cf. *Jasione*   p   * *J. montana*   p

Taxa recorded in cold stage sediments, in Flora Europaea *order* (*cont.*)

169. COMPOSITAE
Compositae   m,p
Compositae less Comp. Lig.   p
Compositae Liguliflorae   p
Compositae Tubuliflorae   p
cf. Compositae   m

* *Eupatorium cannabinum*   m

Solidago t   p

* *Bellis perennis*   m
cf. *B. perennis*   m

Aster t   p
Bidens   p

* *Aster tripolium*   m
* *B. tripartita*   m
* *B. cernua*   m

Anthemis t   p
Achillea   m
cf. *Achillea*   m
Achillea t   p

* *Anthemis cotula*   m
* *A. ptarmica*   m
cf. *A. ptarmica*   m
* *A. millefolium*   m
cf. *A. millefolium*   m

Matricaria t   p
cf. *Matricaria*   m

* *M. maritima*   m
M. cf. *maritima*   m
* *M. perforata*   m
M. cf. *perforata*   m

cf. *Tanacetum*   m

* *T. vulgare*   m
T. cf. *vulgare*   m
* *Leucanthemum vulgare*   m

Artemisia   m,p
cf. *Artemisia*   m

A. vulgaris t   p

cf. *Tussilago farfara*   m

Senecio t   p

S. cf. *jacobaea*   m
* *S. aquaticus*   m

Arctium t   p
cf. *Saussurea*   p
Carduus   m,p

* *S. alpina*   p
* *C. nutans*   m
C. cf. *nutans*   m
* *C. pycnocephalus*   m

Cirsium   m,p
Cirsium t   p
cf. *Cirsium*   m

* *C. vulgare*   m
C. vulgare t   m
* *C. helenioides*   m
# *C. palustre*   m
* *C. arvense*   m

Carduus/Cirsium   p
Carduus/Cirsium spp.   m
cf. *Carduus/Cirsium*   m
Centaurea   m

* *C. scabiosa*   m,p
* *C. nigra*   p
C. nigra t   p
C. cf. *nigra*   p
* *C. cyanus*   p

Leontodon   m
cf. *Leontodon*   m

* *L. autumnalis*   m
L. cf. *autumnalis*   m

cf. *Leontodon* spp.   m

cf. *Picris*   m
Sonchus   m
cf. *Sonchus*   m

Taraxacum   m
Taraxacum spp.   m
Taraxacum t   p
Taraxacum sect. *Spectabilia*   m
Taraxacum sect. *Palustria*   m
Taraxacum sect. *Alpina*   m
Taraxacum sect.
    Erythrosperma   m
Taraxacum sect. *Taraxacum*   m

Crepis   m
cf. *Crepis*   m
Hieracium   m
cf. *Hieracium*   m

cf. *L. autumnalis*   m
L. hispidus   m
L. cf. *hispidus*   m
* *P. hieracioides*   m
* *S. asper*   m
* *S. oleraceus*   m
* *S. arvensis*   m
S. cf. *arvensis*   m
+ *T. officinale* agg.   m

* *Lapsana communis*   m
* *C. capillaris*   m

MONOCOTYLEDONES
Monocotyledones   m

170. ALISMATACEAE
    Alismataceae   m
    cf. *Sagittaria*   p

    Alisma   m,p
    Alisma t   p
    cf. *Alisma*   m

* *Sagittaria sagittifolia*   m
* *S. natans*   m
* *A. plantago-aquatica*   m

* *Damasonium alisma*   m

171. BUTOMACEAE
    Butomus   p

* *B. umbellatus*   m

172. HYDROCHARITACEAE
    Hydrocharis   p

* *H. morsus-ranae*   m
* *Stratiotes aloides*   m

173. SCHEUCHZERIACEAE

* *Scheuchzeria palustris*   m

175. JUNCAGINACEAE
    cf. *Triglochin*   p

* *T. maritima*   m

177. POTAMOGETONACEAE
    Potamogeton   m,p
    Potamogeton spp.   m
    Potamogeton t   p

* *P. natans*   m
P. cf. *natans*
* *P. polygonifolius*   m

Taxa recorded in cold stage sediments, in Flora Europaea *order (cont.)*

Potamogeton	* *P. coloratus* m
subg. *Potamogeton* p	*P.* cf. *nodosus* m
	P. cf. *lucens* m
	* *P.* ×*zizii* m
	* *P. gramineus* m
	P. cf. *gramineus* m
	* *P. alpinus* m
	P. cf. *alpinus* m
	* *P. praelongus* m
	P. cf. *praelongus* m
	* *P. perfoliatus* m
	P. cf. *perfoliatus* m
	* *P. friesii* m
	P. cf. *friesii* m
	* *P. pusillus* m
	* *P. obtusifolius* m
	P. cf. *obtusifolius* m
	cf. *P. obtusifolius* m
	* *P. berchtoldii* m
	P. cf. *berchtoldii* m
	* *P. trichoides* m
	* *P. compressus* m
	* *P. acutifolius* m
	P. cf. *acutifolius* m
	* *P. crispus* m
	* *P. filiformis* m,p
	P. cf. *filiformis* m
	cf. *P. filiformis* m
	* *P. vaginatus* m
	P. cf. *vaginatus* m
	* *P. pectinatus* m
	P. cf. *pectinatus* m
	* *Groenlandia densa* m
181. ZANNICHELLIACEAE	
	* *Zannichellia palustris* m
182. NAJADACEAE	
	* *Najas marina* m
	* *Najas flexilis* m
	Najas cf. *flexilis* m
183. LILIACEAE	
Liliaceae p	
cf. *Veratrum* p	
	cf. *Lloydia serotina* p
cf. *Fritillaria* p	
Allium m,p	* *A. schoenoprasum* m
Allium t p	
188. IRIDACEAE	
Iris p	* *Iris pseudacorus* m

189. JUNCACEAE
    Juncus m
    Juncus spp.
    Juncus spp. incl. *J*. cf.
        effusus/inflexus m
    Juncus spp. incl. *J. bufonius* m
    cf. *Juncus* m

    J. cf. *maritimus* m
  * *J. balticus* m
    J. cf. *balticus* m
  * *J. inflexus* m
  * *J. effusus* m
    J. cf. *effusus* m
    J. effusus and/or *J. conglomeratus* m
  * *J. conglomeratus* m
    J. cf. *conglomeratus* m
    J. conglomeratus t m
    cf. *J. conglomeratus* t m
  * *J. gerardii* m
    J cf. *gerardii* m
  * *J. bufonius* m
  + *J. bufonius* agg. m
    J. cf. *bufonius* m
    cf. *J. bufonius* m
  * *J. subnodulosus* m
  * *J. bulbosus* m
    J. bulbosus t m
  * *J. acutiflorus* m
    J. articulatus t m
    J. articulatus/acutiflorus m
    J. cf. *triglumis* t m

    Luzula m
    Luzula spp. m
    cf. *Luzula* m
    Luzula? m

    L. cf. *multiflora* m
  * *L. spicata* m
    L. cf. *spicata* m

193. GRAMINEAE
    Gramineae m,p
    cf. Gramineae m
    Gramineae Festucoid t m
    Festuca m
    Festuca spp. m
    cf. *Festuca* m

  * *F. rubra* m
    F. cf. *rubra* m
    F. rubra? m
  * *F. halleri* m
    F. tenuifolia? m
  # *F. ovina* m
    F. ovina? m
    cf. *P. annua* m
    P. cf. *trivialis* m
    P. cf. *pratensis* m

    cf. *Poa* m
    Agrostis/Poa t m
    Poa spp. *and/or Elymus* spp. m
    Glyceria m
    Glyceria t p
    cf. *Glyceria* m

    G. cf. *declinata* m
    cf. *G. fluitans* m

    cf. *Hordeum* and/or *Elymus* m

    cf. *Elymus repens* m

  * *Anthoxanthum odoratum* m

    Agrostis m

Taxa recorded in cold stage sediments, in Flora Europaea *order* (*cont.*)

cf. *Agrostis* m
Alopecurus m

196. LEMNACEAE
    Lemna m,p
    cf. *Lemna* p

    * *L. trisulca* m
    L. cf. *trisulca* m
    # *L. minor* m

197. SPARGANIACEAE
    Sparganium m,p
    Sparganium spp. m
    Sparganium t p
    cf. *Sparganium* m

    * *S. erectum* m
    ** *S. erectum* ssp. *neglectum* m
    S. erectum t p
    * *S. emersum* m
    S. cf. *emersum* m
    * *S. angustifolium* m
    S. cf. *angustifolium* m
    * *S. minimum* m

198. TYPHACEAE
    Typha m,p
    Typha spp. m

    * *T. angustifolia* m
    T. angustifolia t p
    * *T. latifolia* m,p
    T. latifolia t m
    T. cf. *latifolia* m

199. CYPERACEAE
    Cyperaceae m,p
    Cyperaceae? m
    Scirpus m
    Scirpus spp. m
    cf. *Scirpus* m

    * *S. lacustris* m
    S. cf. *lacustris* m
    cf. *S. lacustris* m
    ** *S. lacustris* ssp. *lacustris* m
    ** *S. lacustris* ssp. *tabernaemontani* m
    S. cf. *lacustris* ssp. *tabernaemontani* m
    S. lacustris ? m
    * *S.* × *carinatus* m
    S. cf. *pungens* m
    S. cf. *americanus* m
    * *S. setaceus* m
    * *Blysmus compressus* m
    * *Blysmus rufus* m
    Eriophorum m
    * *E. angustifolium* m
    * *E. vaginatum* m
    Eleocharis m
    * *E. quinqueflora* m
    Eleocharis spp. m
    * *E. parvula* m
    cf. *Eleocharis* m
    * *E. palustris* m
    + *E. palustris* agg. m
    E. cf. *palustris* m
    cf. *E. palustris* m
    ** *E. palustris* ssp. *palustris* m

cf. *Cyperus* m
Cladium p
Rhynchospora? m
Carex m
Carex spp. m
Carex-biconvex m
Carex-trigonous m
Carex sect. *Acutae* CTW m
cf. *Carex* m

** *E. palustris* ssp. *vulgaris* m
E. palustris + *E. uniglumis* m
* *E. uniglumis* m
E. cf. *uniglumis* m
* *E. multicaulis* m
* *E. carniolica* m
* *Cyperus longus* m
* *C. mariscus* m
* *R. alba* p
* *C. paniculata* m
* *C. appropinquata* m
C. cf. *diandra* m
* *C. divisa* m
* *C. maritima* m
* *C. ovalis* m
* *C. dioica* m
* *C. lachenalii* m
C. lachenalii? m
C. cf. *curta* m
* *C. hirta* m
* *C. acutiformis* m
C. cf. *acutiformis* m
* *C. riparia* m
C. cf. *riparia* m
C. cf. *pseudocyperus* m
* *C. rostrata* m
C. cf. *rostrata* m
C. rostrata t m
cf. *C. rostrata* m
* *C. vesicaria* m
C. cf. *vesicaria* m
* *C. pendula* m
C. cf. *pendula* m
* *C. capillaris* m
* *C. strigosa* m
* *C. flacca* m
C. flacca? m
* *C. panicea* m
C. cf. *panicea* m
* *C. distans* m
* *C. punctata* m
* *C. flava* m
+ *C. flava* agg. m
* *C. pallescens* m
* *C. atrata* m
C. cf. *recta* m
C. aquatilis m
C. cf. *aquatilis* m
C. cf. *aquatilis/bigelowii* m
C. bigelowii/aquatilis m
C. bigelowii m
C. bigelowii t m

Taxa recorded in cold stage sediments, in Flora Europaea *order (cont.)*

C. cf. *bigelowii*   m
* C. *nigra*   m
C. *nigra* t   m
C. cf. *nigra*   m
C. *nigra* group J&T   m
C. cf. *nigra* group J&T   m
C. cf. *nigra* t   m
* C. *acuta*   m
C. cf. *acuta*   m
C. cf. *pauciflora*   m
* C. *capitata*   m
* C. *pulicaris*   m
C. cf. *pulicaris*   m

OTHER TAXA
BRYOPHYTA
Sphagnaceae
   Sphagnum spp.   p

CHARACEAE
Characeae   m
   Chara   m
   Chara spp.   m
   cf. *Chara*   m
   Nitella   m
   Nitella t   m
   Tolypella   m   *T. nidifica*   m
                                               cf. *T. nidifica*   m

MISCELLANEOUS
   Type x   p
   Nyssa   p
   Pettitia   m

   Sphacelaria cf. *plumigera*   m

   Megaspores   m
   Reworked Quaternary pollen   p
   Pre-Quaternary palynomorphs   p

Index

References to Chapter 7, The Flora, are marked by a tick ('); they are confined to families and genera, making reference to species readily available.

The taxa list in Appendix II is indexed for families only; these entries are marked by a double tick (").

Major consideration of plant species and Paradox tables is distinguished by bold type.

References to tables and figures are marked with an asterisk (*).

Topics and authors are indexed only for the more important references.

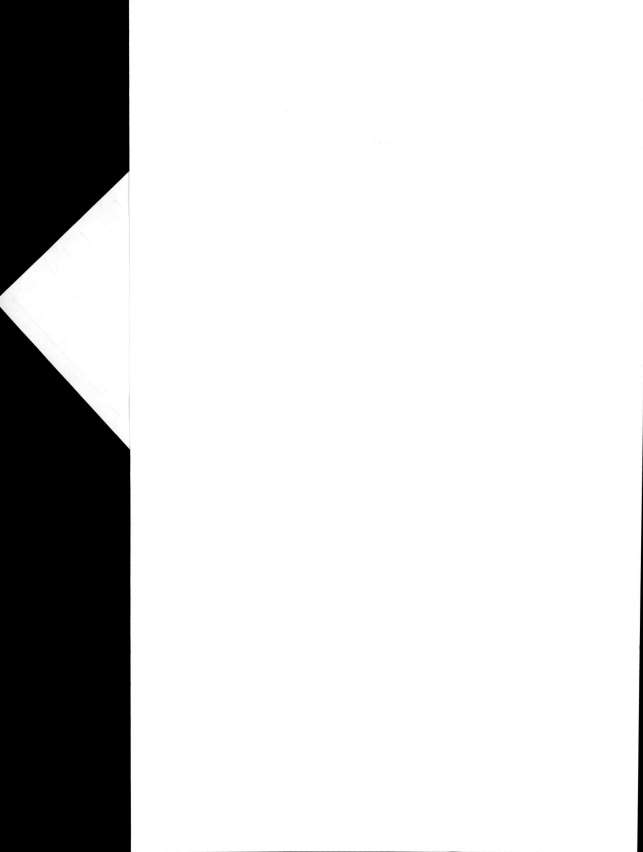

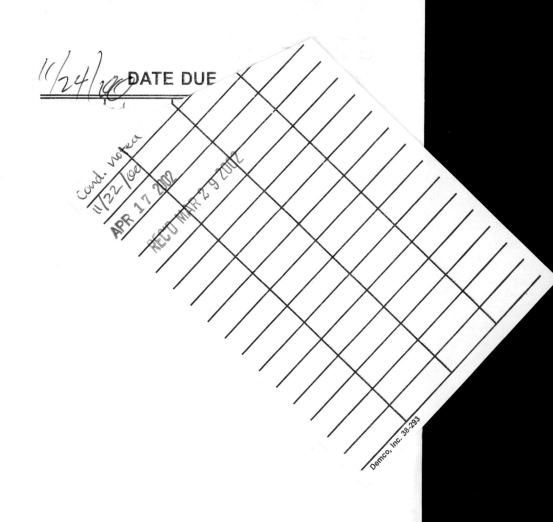

11/24/00

DATE DUE

Cond. notes
11/22/00

APR 17 2002

REC'D MAR 29 2002

Demco, Inc. 38-293